Biobanks in Healthcare

Series Editor
Nicole Arrighi

Biobanks in Healthcare

From the Collection of Biological Samples to Digital Health

Nicole Arrighi
Paul Hofman

iSTE

WILEY

First published 2024 in Great Britain and the United States by ISTE Ltd and John Wiley & Sons, Inc.

ISTE Ltd
27-37 St George's Road
London SW19 4EU
UK

www.iste.co.uk

John Wiley & Sons, Inc.
111 River Street
Hoboken, NJ 07030
USA

www.wiley.com

Library of Congress Control Number: 2024943915

British Library Cataloguing-in-Publication Data
A CIP record for this book is available from the British Library
ISBN 978-1-78630-023-2

Contents

Foreword

If we go back as far as possible in the history of humanity, it is possible that the first gesture of the first researcher was to collect objects taken from nature. Gathering, classifying and preserving these objects in order to observe and study them were probably at the origin of the first scientific approach of our ancestor, gifted with consciousness and abstraction, and the irrepressible desire to know the world. More recently, in the 17th-century curiosity cabinets, it was a question of enthralling the visitor, but also of affirming one's rank and power. European kings understood this well when they created museums and other royal gardens that brought together collections of extraordinary animals and plants taken from the confines of their kingdom. These collections were at the origin of the change in our knowledge of life sciences. Just think of Buffon's monumental *Natural History* during the Enlightenment!

Human biological collections were established with the advent of hospitals in the 19th century, which became not only places of charity but also establishments for care and for the transmission of knowledge. In order to describe and try to understand diseases or congenital anomalies, it was necessary to collect, preserve and classify affected organs and fragments of biological tissues. Thus, the famous anatomical collections of Jean-Martin Charcot at the Salpêtrière Hospital, of Marc-Antoine Petit, the first surgeon-major of the Hôtel-Dieu Hospital in Lyon, and of the medical schools of Montpellier and Strasbourg, were created. This period also gave birth to anatomopathology, the study and classification of the macroscopic and microscopic lesions of injured tissues, with Jean-Baptiste Morgagni in Padua, Rudolf Virchow in Berlin and Jean Lobstein in Strasbourg. Significant collections were thus accumulated over decades in accordance to the passions of these physician-researchers; collections that most often ended up in the dark basements of hospitals some time after the disappearance of their creators, closely followed by that of the notebook blackened with cabalistic annotations understandable only by its author.

At the dawn of the 21st century, with the development of high-throughput analysis techniques, access to quality biological samples became a major issue. Indeed, it quickly became apparent that the growing industrialization of research in biology and health required that biological objects be of impeccable quality and provided with reliable annotations for analysis. The first databases resulting from the exploitation of human samples were often marred by errors and generated erroneous scientific information. Another major issue was that of economic value. The development of biotechnologies, the rise of biological drugs and the identification of biomarkers associated with molecular classifications of tissue lesions made it essential to have orderly access to human biological samples, which had become precious and rare.

France was a pioneer at the international level in organizing and defining the missions of biological resource centers, or biobanks, new biology–health infrastructures designed to collect, store and make available human biological samples. It was also the first country to establish a standard for the operation of biobanks, which was recently converted into an ISO standard, and to establish ethical rules for access to biological samples from the human body, which are included in the Public Health Code. The normative progress has been considerable, the scientific production improved and the services rendered to the advancement of medicine indisputable. Functional biobanks today are those that combine quality assurance of biological samples and their annotations, technical and scientific expertise, transparency of governance, innovation, training and integration into the local, national and international research community.

More than 20 years after their creation, the biobanking project is not yet complete [CLÉ 19]. The business model is still unstable due to the lack of a real national strategy to structure and finance a national network of biobanks based on independent peer evaluations. In many cases, stocks are thus increasing indefinitely, in the hope of a hypothetical valorization. Moreover, access to epidemiological, bioclinical and genomic data associated with collections remains difficult, in the absence of harmonization of hospital information systems and a real willingness to share these data. Despite important advances at the European level, the legislative and regulatory framework is slow to adapt to the evolution of biobanks and the constitution of large cohorts that involve the storage of biological samples and associated data for future research.

Biobanks are at a crossroads. Like all essential infrastructures in biology and health, they must adapt to advances in research and technology. In the digital era and with the production of Big Data by practically all laboratories that have access to very high-throughput and low-cost technological platforms, biobanks are becoming the guarantors of the reproducibility of experiments and of the overall quality of research in biology and health. They are expanding their scope of intervention, automating processes and exploring new analysis techniques with precious samples, while preserving them for future generations. They are diversifying their storage capacities; in addition to information on biological objects, epidemiological, clinical, biological, genomic and imaging data are being included in their computer bases. Biobanks, by their very nature, are interface infrastructures, which erase the artificial divide between care and research activities. A striking example is their central role in medical genomics, which relies on the intensive deployment of digital technology to produce information that is simultaneously useful for patient care and for researchers studying disease.

This book reviews the current state of biobanking. It explores new avenues, in particular the change linked to the massive deployment of digital health and precision medicine. The authors show that the future of biology and health requires the deployment of biobanks in fields that have yet to be explored, at the forefront of the extraordinary adventure of 21st-century research.

Bruno CLÉMENT
Director of research at Inserm
Institute of Nutrition, Metabolisms and Cancer
CRB Santé Biological Resource Center of Rennes University Hospital

Acknowledgments

This book is the result of collegial research that contributed to the creation of the Master of Science in Biobanking and Complex Data Management at Côte d'Azur University. The authors would like to thank all the experts from the various European and North American countries who have contributed to and continue to enrich the training program for future biobank managers.

Dr. Bruno Clément has given us the honor of writing the foreword to this book. He is the director of the Inserm U991 unit "Liver, Metabolism and Cancer" at the Pontchaillou Hospital of the Rennes University Hospital. Always extremely involved in action in favor of innovations in biotechnology, he strongly participated in the creation of the national project for biobanks and has played an essential role in the IBISA infrastructures (platforms of resources in biology, health and agronomy open to the scientific community).

Dr. Robert Hewitt, founder of the Biosample Hub and co-founder of the European, Middle Eastern and African Society for Biopreservation and Biobanking (ESBB), played a central role in the construction of the course by sharing his expertise in biobank management and his experience in the industry. We would like to thank him for his presence at the inauguration of the defense juries for the Master of Science in Biobanking and Complex Data Management at Côte d'Azur University.

Dr. Marius Ilié is a pathologist and a professor in the Faculty of Medicine of Nice and a member of the biobank scientific council of the laboratory of clinical and experimental pathology at Pasteur University Hospital in Nice.

Through his expertise in sample management, he has contributed to numerous publications in biomedical research. He is a member of IRCAN team 4 (Inserm U1081/UMR CNRS 7284).

Dr. Élodie Long-Mira is a physician of molecular and clinical pathologies. Her research activities at the laboratory of clinical and experimental pathology of the Pasteur University Hospital in Nice are focused on lung cancer and melanoma. She is a lecturer at the Faculty of Medicine in Nice and teaches embryology and histology. She is a member of IRCAN team 4 (Inserm U1081/UMR CNRS 7284) and in charge of the melanoma collection at the Nice biobank.

Dr. Giorgio Stanta is Professor of Pathology at the University of Trieste. His major interest is in the application of molecular analysis to formalin-fixed and paraffin-embedded tissue samples. He has demonstrated the importance of pre-analytical conditions and standardization of genomic and proteomic methods. He is an expert of the BBMRI-ERIC network and coordinates the European group IMPACTS Archive Tissues: Improving Molecular Medicine Research and Clinical Practice[1].

Dr. Georges Dagher was the Director of the French Biobank Infrastructure and Director of Research at Inserm (France). In addition to his clinical research activities, he has contributed to the OECD guidelines on good practices for biological resource centers, human biobanks and genetic databases. He participated in the development of the French standard NF S96-900 and its transformation into an international ISO standard for biobanks.

Dr. Kurt Zatloukal is Director of the Austrian Biobanking and Biomolecular Resources Research Infrastructure (BBMRI). He is a pathologist and professor at the Medical University of Graz (Austria) and heads the Christian Doppler Laboratory for biospecimen research and biobanking technologies. As coordinator of the preparatory phase of the BBMRI infrastructure, he is involved in the development of new European standards in the pre-analytical phase of molecular diagnosis of tissue samples.

1 www.impactsnetwork.eu.

Taycir Skhiri is project manager of the FHU OncoAge (University Hospital Federation Oncoage, www.oncoage.org) the goal of which is to share research results and develop innovative solutions in the FHU OncoAge. Converting and applying the results of research requires the transmission of knowledge that meets, among other things, the demand for the professionalization of biobanks.

Dr. Kevin Washetine is a quality engineer in the clinical and experimental pathology laboratory of the Pasteur University Hospital in Nice. He coordinates the implementation of quality procedures. His work has enabled the laboratory to obtain and maintain certification according to the NFS 96-900 standard (quality of biological resource centers) from AFNOR and accreditation by COFRAC according to the ISO 15189 standard for pathology and molecular biology activities.

Zeineb Messaoudi is a quality engineer at the clinical and experimental pathology laboratory of the Pasteur University Hospital in Nice. She is implementing the transition from the NFS 96-900 standard to the international standard for biobanks ISO 20387 (quality management system for all biobanking activities).

Dr. Laurent Counillon is Director of the EUR Life University Research School, the infrastructure that, among other things, pilots the IDEX Masters of Science (initiatives of excellence launched by the French government). These interdisciplinary programs, taught at Côte d'Azur University in English, focus on creating programs of excellence with international visibility.

Dr. Sophie Raisin-Tani is a professor at Côte d'Azur University. Her involvement from the beginning of the project has accelerated the sharing of resources and the opening of the project to the international scene, increasing its attractiveness to students and biobank managers invited to participate in the teaching.

We wish to thank all the members of FHU OncoAge who, thanks to their multiple translational research activities, actively contribute to the influence of the Nice University Hospital biobank (BB-0033-00025).

Introduction

Biobanks in 2022: Why and How?

Research is based on several foundations: fundamental concepts, cellular and animal models, validation of discoveries obtained in laboratories through a "translational" approach and then, depending on the sector of interest, clinical or applied research or research carried out *in silico* using databases. Depending on the research topics, it is necessary to achieve a continuum between the discovery of new cellular mechanisms and an application in the living world, whether it is the plant, microbial or animal and human world.

One of the unavoidable links between basic research and its various applications is the use of different biological resources, whether they come from plants, microbes, animals or healthy or sick human beings [ALD 19, VAU 19]. From time immemorial, researchers have needed to analyze these different biological resources to validate their hypotheses or observations made with cellular models. Several purposes have progressively emerged, such as better understanding of developmental biology or of the mechanisms of cell death, growth or transformation. One of the most successful examples of the use of biological resources is the discovery of biomarkers of human diseases, which can be biomarkers for diagnostic, prognostic or predictive purposes in a therapeutic response [BAR 20, HEW 11]. While the these biological resources were initially used in a poorly controlled manner and without any real established rules, several reflection policies have gradually led to the establishment of a controlled and rigorous operation. Thus, the notion of a *collection* was born, associated with the need to control use and the necessary storage spaces. In recent decades, these different collections have been associated with the creation of *biological resource centers* (BRCs), still called "biobanks" today, the structuring of which has been progressive, and they have become essential "tools" in the world of research [WAT 19].

Biobanks have evolved in recent years and are now professionalized and complex structures whose operation is subject to national and international regulations. These biobanks can function as secure storage areas for biological collections (called "biorepositories"), making samples available for research projects. They can also be organized as *Expert Centers*, and thus both offer services to researchers (thanks to the existence of technical platforms) and enable scientific collaboration (thanks to unique knowledge). Biobanks can also develop their own research projects on specific topics, in particular on the control, management and use of biological samples, essentially with the aim of optimizing their operations and their service offer [WAS 18a].

Biobanks and collection policy have evolved over time. A typical example is the evolution of biobanks collecting patient samples. At first, many samples were accumulated and frozen, in particular tumor samples. The question of the quality of these thousands of stored samples then arose. Indeed, the discovery and/or validation of the different markers proved to be totally dependent on the reproducibility of the analyses on perfectly preserved samples. Finally, the need to acquire more and more associated or targeted data modified the functioning of biobanks by requiring the development of multiple and complex databases. Indeed, one of the key points associated with making biological samples available for research projects is now the need to associate very precise and relevant data with them. The world of biobanking has therefore evolved progressively over the last few decades, with this field becoming a *medical specialty in its own right*. Thus, schematically, the activity of biobanks has passed through several stages, from the "biobank 1.0" (massive collection activity often without any preconceived ideas and based on "quantitative" data) to the "biobank 2.0" (where the quality of samples is considered crucial along with the implementation of the control of the parameters of the pre-analytical phase), and finally the "biobank 3.0" (integrating several imperatives, in particular the controlled management of clinicobiological data, the anticipation of requests and the control of the business model) [DOU 17, ELL 15, LIN 20, SIM 14]. Thus, the creation and construction of a biobank-type structure must nowadays meet strict requirements, in particular associated with a collection strategy and planning of the use of stored samples [BAI 16, HOF 13].

The diversity of expertise required to generate the data associated with the samples, their integration and their analysis necessitate a rethinking of the organization and structuring of biobanks in order to address the services of specialists in different fields, notably medical, biological, imaging, statistics, bioinformatics and mathematics. This change must concern, at the national or international level, biobanks involved in the collection, integration and analysis of complex data. The concept of next generation biobanks (NGBs) is gradually emerging. An NGB must be able to combine huge databases hosting phenotypic, behavioral, familial, imaging, omics, radiomics and biological data from different centers. Operating an NGB requires overcoming several challenges, the first of which is that of computing infrastructures capable of processing tetrabits, pentabits and exabits. In addition, a suite of appropriate analysis methods and algorithms must be developed. Moreover, these NGBs are now accompanied by a paradigm shift from the analysis of data from a large number of patients to the analysis of a large volume of data from a single subject ("Big Data" vs. "Fat Data"). A crucial aspect for biobanks is their sustainability, taking into account the fact that the business model of these structures is often fragile [RAO 19, VAU 11]. Indeed, the operation of a biobank today requires a dedicated and qualified staff, a secure and expensive infrastructure and equipment that must be renewed regularly in order to maintain the quality of the collections.

There are more and more biobanks throughout the world and the number of collections that can be made available to researchers is increasing, which may lead to a gradual decrease in demand for certain biobanks due to competition. In this context, in order to be competitive, a biobank must make organizational and strategic choices. It is difficult to maintain a high level of expertise in several fields of activity, and focusing on a limited number of pathologies certainly makes it possible to concentrate on the completeness of the collections and to associate complete clinical and biological data, the latter becoming increasingly complex.

An essential point is to obtain the signed *consent* of patients to use their sample for research purposes. This key point must be associated with an internal process specific to each biobank, which must be perfectly mastered and conducted in consultation with the clinical services. The external *visibility* of a biobank and its recognition are increased when the partners of a biobank have access to different forms of expertise (e.g. histological diagnosis performed by senior pathologists in the field of pathology concerned; expertise in molecular biology and accessible genomic databases). Focusing on one or a few pathologies also makes it possible to associate, for the same patients, diversified collections of tissues (fixed and frozen) and biofluids (plasma, sera, buffy coat, whole blood, urine, other fluids, etc.). Clinical data and in particular the follow-up of patients according to the different successive treatments or the collection of events (progression, metastasis, death) can be more easily integrated into the biological databases [WAT 17]. In this context, it is often easier to join a national or international network of experts in the same field.

Certification or *accreditation* of the biobank according to national or international standards is also essential to the robustness of its operation and the quality of the service provided. A competitive biobank must also develop innovative projects, particularly in relation to the pharmaceutical and biotechnology industries. Different projects can allow for the transfer of the results of the innovation thus acquired to clinical practice, after a validation phase carried out with biological samples. Thus, the visibility of a biobank vis-à-vis different partners and/or applicants for biological samples can benefit from the implementation of *performance indicators* that must be adapted to the structure concerned and its ambitions [HOF 13].

Several challenges are emerging in the near future for biobanks, and these structures must meet a certain number of common objectives. For example, data from different databases or documents (pathological anatomy, molecular biology, imaging, clinical and therapeutic data) have to be gathered (or used) in a single accessible source. It will also be necessary to integrate, store and process complex and often very heterogeneous data volumes. One of the challenges is the sharing and access to information, which must be secure and based on perfectly de-identified data. This access could be envisaged for external partners wishing to know the available samples and the associated complex data. In view of these rapid developments and a change in the activities related to the biobanking profession, it is therefore crucial to continue or develop appropriate university training and to recruit people into the biobanks who have benefited from practical and theoretical training in this field. The control (capture and exploitation) of complex and Big Data associated with biological samples collected from cancer patients has become a major issue in oncology. This control requires several combined actions. The first step is to control the *quality of the data* collected. In the world of Big Data, there is a major risk of generating secondary data biased by incomplete or false primary data. No matter how sophisticated the tools deployed, especially those based on deep learning, machine learning and artificial intelligence approaches, it is impossible to turn "lead into gold", i.e., to generate quality data from insufficient or erroneous initial data. The pathologist's role is essential here and only their expertise can verify the quality of the diagnosis and the analyses performed on the tissue sections. The integration of associated molecular biology data is also a key step. These data are of varying complexity and may include whole genome sequencing data for a tumor, and also transcriptomic, proteomic and metabolomic data. Once again, the quality of the data (also related to the quality of the samples and therefore of the pre-analytical steps) is the essential point to control.

Biobanks are subject to strict *legislation* and the use of samples for research projects can only be undertaken in compliance with the law. Thus, public–public or public–private partnerships can only be established through contracts and signed procedures managing material transfer agreements (MTAs) [CLÉ 14, HOF 14, SIM 20]. These contracts must specify the purpose of the research carried out with the samples and the identity of the end-user. Indeed, some private companies serve as platforms for the recruitment of biological samples, especially from academic laboratories, and supply users without knowing the purpose of their research.

The biobank world is not immune to the arrival of *artificial intelligence* (AI) [LIN 20, MAR 20, NAR 20]. AI is in fact a very general term to designate "the set of theories and techniques used to create machines capable of simulating intelligence". Often classified as one of the cognitive sciences, it uses computational neurobiology, mathematical logic and computer science. Since its birth 70 years ago with a few simple abstractions of how a neuron works in the human brain, AI has become an indispensable tool for computer vision applications. In particular, *artificial neural networks* (ANNs) have gained great popularity because of their impressive recognition capability when implemented with many layers of artificial neurons (processing units that can perform simple aggregation of incoming synaptic values from other units). Given the amount of data integrated in a biobank, the use of artificial intelligence software to select collections based on complex data, including patient follow-up according to treatments, seems unavoidable in the future. However, it may seem absurd to continue to collect samples when biobanks are currently almost saturated and the demands only allow for a very low level of destocking in general compared to the available stocks [RUS 19].

One might also think that there is so much omic data available, especially on public databases, that it will be less and less useful to accumulate the different biological products. In reality, this does not take into account evolution and climatic and environmental changes and their impact on the emergence of human diseases. One of the recent examples concerns the collection of biological samples from patients with *Covid-19*. Several collections have thus been rapidly formed, thanks to already existing and efficient infrastructures allowing the organization of different collections of blood or cellular products associated with clinicobiological databases [HUA 20, PEE 20, ROV 20, TAN 20]. These collections have sometimes been established prospectively and supported by research projects. Thus, it is thanks to the biobanks that the management of these new collections could be perfectly controlled according to the sanitary and personnel protection rules, ensuring the traceability and quality of the samples for future users. However, certain specific situations, such as the appearance of infectious diseases or economic crises, can have serious consequences on the sustainability of a biobank. Indeed, it is also possible, given the impact of an emerging infection on the functioning of hospitals, that surgical interventions or hospitalizations of non-infected patients suddenly decrease, thus limiting the flow of samples to biological and pathology services and thus the possibilities of collections [HOF 20, VAU 20].

The future of biobanks is linked to the evolution of technologies, especially from the biomedical world, and also to knowledge from other disciplines such as the social and mathematical sciences [HAR 19, KIN 20]. It is very likely that increasingly complex databases will lead to limitations in the collection and storage of biological samples. “Infobesity” is leading to the progressive saturation of computer storage spaces with these different massive data, their security cost is increasing and a lot of information could in fact be kept within the nucleic acids managed by biobanks. Finally, one of the challenges to be met in the future will be for the different biobanks to be able to work together throughout the world on common subjects, which requires ensuring the same criteria for the quality and accessibility of data are used in order to facilitate the flow of sample exchanges and to allow ambitious research projects [VAU 14].

1

Biobanks, a Source of Human Samples and Health Data

Precision medicine is forcing modern biobanks to shift their focus from sample-driven to data-driven strategies.

[KOZ 19]

In October 2018, the scientific journal *Nature* provided an international platform for the UK Biobank, a biobank of biological samples. This biobank collected biological samples from more than half a million individuals between 2006 and 2010. All volunteers and spread throughout the United Kingdom, 500,000 individuals gave their time, their medical data and their samples.

In this issue of *Nature*, and for the first time, the publications gave a description of the entire cohort, including the genetic data of all individuals [BYC 18]. Then, in a second study, the researchers showed the brain imaging results of 10,000 individuals. They revealed the genetic influence on brain structure and function and showed correlations between personality and neurodegenerative and psychiatric traits [ELL 18]. These results are valuable because they contribute to the study of a population cohort and will be used in other genomic medicine projects.

The UK Biobank is the first project to carry out a large-scale collection. It successfully demonstrates, on a population scale, that clinical, genetic and physical information are linked. Other countries have also launched biobanking projects such as Estonia, Japan, Canada, Finland and Iceland. However, the UK has overtaken them. In 2010, the UK Biobank was building a prospective cohort of some 500,000 individuals aged 40–69 years.

In the October 2018 *Nature* publications, Bycroft [BYC 18], Elliot [ELL 18] and their teams described an immense resource of genomic data linked to clinical data from 500,000 individuals. This large-scale biobank promises to uncover relationships between genomic variations and diseases and help better understand their mechanisms.

As described by Bycroft et al., the 500,000 UK Biobank participants gave urine, saliva and blood samples, which were used for genetic analysis and *biomarker* testing of known diseases. Participants were between the ages of 40 and 69 between 2006 and 2010 when they were included in the study. This age range corresponds to the age of onset of disease in adults. Volunteers filled out questionnaires that included information on several factors, such as family history of disease and lifestyle. They also gave the researchers access to their medical records.

While Bycroft and his team detailed the generation of genomic data from the biobank, Elliot et al. provided insight into how these data could help discover genetic associations of disease. The authors explored brain imaging data from more than 8,400 participants. These data were processed to generate a list of thousands of image-derived phenotypes (IDPs), which are features related to brain structure and function that can be identified from medical imaging images. Elliot and his team studied the associations between IDPs and genetic variants.

In many population-cohort studies, access to data is not possible until the results are published. In some cases, some researchers do not even make all the data available. The UK Biobank, funded by the Medical Research Council and the Wellcome Trust, has made the complete data available to researchers.

The size of this resource is unprecedented in terms of the number of samples and the volume of data collected. From the beginning, the premise of the project was based on sharing all data with any human health researcher or scientist. As a result, thousands of scientists around the world have been able to conduct their research projects on this data since July 2017.

Is the UK Biobank unique? Are there other biobanks? What are their activities?

In this first chapter, we define the concept of the human sample bank by providing examples of population and clinical biobanks. Then, we present their mapping according to geographical distribution by locating the major international biobanks. Finally, we describe the management of the biological samples and processes of a biobank, which combine technical expertise and the infallible maintenance of the quality of the human sample within a legal and ethical framework.

1.1. From the collection of biological samples to the concept of a biobank

According to the principles of data sharing and access to samples for all researchers, biobanks must be equipped with communication tools to facilitate the exchange of material and knowledge. The work of Fransson et al. has highlighted the priority of a common language shared among all actors in the multidisciplinary field of biobanking [FRA 15].

Biobanking activities mobilize the resources of biology to describe the properties of samples, of medicine for clinicopathological annotations, of informatics to manage the data associated with the samples and of law to define the regulatory framework for sample donation and to protect the personal data of patients. The terminology used by the biobanking community is found in the lexicons of biology, medicine and law. A preliminary explanation of the words used by biobank users and scientists using the associated data is a necessary preamble to entering the biobanking world.

Fransson et al. analyzed the language used in biobanks and showed that there is no universal definition of the word "biobank". They selected a dozen terms related to biological sample management activities, which they submitted to biobank users. The co-authors – an epidemiologist, a lawyer and a computer scientist – asked about 100 people working in European biobanks to choose the most appropriate definition from a list of proposals

[FRA 15]. This study identified a wide range of definitions. Two answers came out on top in defining a (human) *biobank*.

The first definition explains the concept of a biobank as "an organized collection of human biological material and associated information stored for one or more research purposes". This definition comes from the lexicon of the international consortium called the Public Population Project in Genomics and Society (P3G). This non-profit consortium brings to the international scientific community expertise, resources and innovative tools for medical and social science research.

The second definition comes from the European biobanking infrastructure called the Biobanking and Biomolecular Resources Research Infrastructure – European Research Infrastructure Consortium (BBMRI-ERIC). Biobanks are defined as "collections, repositories and distribution centres of all types of human biological samples, such as blood, tissues, cells or DNA and/or related data such as associated clinical and research data, as well as biomolecular resources, including model- and micro-organisms that might contribute to the understanding of the physiology and diseases of humans". The authors state that two-thirds of the responses came from biobank staff involved in clinical trials, which explains the development of the activities specified in the second definition. The remaining third were from a population biobank.

In this study, Fransson et al. show that some terms overlap, such as sample and specimen. The *sample* is the unit of material derived from a specimen according to the Organization for Economic Cooperation and Development (OECD). The *specimen* is a specific tissue, such as a blood or urine sample, taken from a participant at a specific time. According to the National Cancer Institute's thesaurus, the term *biospecimen* defines any material sample taken from a biological entity for scientific research.

Other terms discussed, such as "sample collection", "aliquot" and "personal data", focus on the research orientation and exclude clinical applications. Ideally, definitions should be universally accepted across sectors to limit confusion among the biobanking community.

The number of publications mentioning the term "biobank" is a reliable indicator of the activity's development. Over the last 10 years, this number has increased by a factor of more than 10 according to the PubMed NCBI bibliographic database. There were 223 publications in 2010 and 2687 in 2020 and 2021. Among the publications of the year 2020, the term "biobanking" was cross-referenced with each of the following terms: health, quality, ethics, economics and informatics. It was found that human health was by far the most important objective of (human) biobanks, followed closely by sample quality and informatics resources. Two disciplines appeared in the ranking with a small but growing number of publications each year: bioethics and economics.

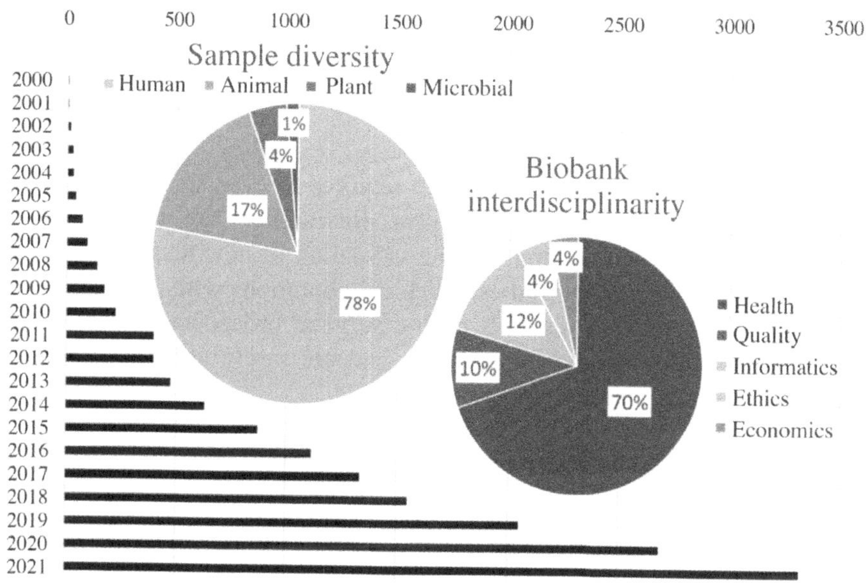

Figure 1.1. *Annual number of publications with the term "biobanking" from 2000 to 2021 according to the PubMed NCBI bibliographic database. For a color version of this figure, see www.iste.co.uk/arrighi/biobanks.zip*

COMMENTARY ON FIGURE 1.1.– *The various biological samples are from human, animal, plant and microbial organisms. In the publications from 2020, the disciplinary fields concerned were health, quality, informatics, ethics and economics.*

Bioethics highlights the fact that in genetic databases the obligations of confidentiality and donor consent are indispensable. The work of Godard et al. raised the issue in 2004 [GOD 04]. They showed how public consultations have raised awareness of sample donation in Iceland, Estonia, the United Kingdom and Quebec, the countries where the first large-scale biobanks were located. They communicated about the ethics of their project, and the researchers gained the trust of the population. Patient volunteers entrusted their personal and medical data to the researchers to build the first cohort-populations. In 2005, Baird and Frome addressed the cost of storing hundreds of samples at very low temperatures [BAI 05]. The investments in infrastructure (building and freezer equipment) and operating costs represent a significant financial commitment, hence the need to design a business model for large-scale biobanks in order to ensure their longevity. Publications on this topic are becoming more frequent.

1.1.1. *The biobank concept*

The development of medical research requires an ever-increasing number of qualified human tissues and samples. Researchers are faced with an eternal need for human material for the validation of new therapeutic tools. They consult with surgeons and establish collaborations with clinicians who will take specimens from patients during surgical procedures. Samples are mainly reserved for diagnostic purposes and are rarely used for scientific research. Access to samples is therefore limited.

Once the sample has been identified and authorized (date and time of the procedure, mode of transport of the sample), the researcher must be able to be sure of its conservation as it guarantees the sample's quality. Part of the tissue or tumor is analyzed and the remaining part is stored at very low temperature or used in another project.

In a hospital department, the storage of samples for a long period of time is impossible due to lack of space. The researcher must therefore diversify their sources of specimens, especially for rare diseases (multicenter cohort). From the need to access samples and considering the investment in time devoted to the search for samples – and not to the experiments – the idea emerged to create a structure specific to the scientific community with:

– hundreds of samples;

– organized collections;

– high-quality specimens and samples.

Thus, the concept of the biobank was born. According to Hewitt and Watson, the term biobank is applied to biological collections of samples of human, animal, plant and microbial origin [HEW 13]. It refers not only to collections of samples and their associated data, but also to the management of samples according to professional standards. The term biobank is general, regardless of the size of the biobank, the number and the purpose of the samples. Generally speaking, biobanks are professional “repositories” of biological samples.

1.1.2. *The first biobank described in the Framingham study*

The first biobank was described in the Framingham Heart Study. Begun in 1947, this epidemiological approach to heart disease was carried out with the population of Framingham, a town located about 30 km from Boston in Massachusetts [DAW 66]. The 28,000 inhabitants of Framingham constituted a representative sample of the American population. The cohort was large, with 5,209 individuals aged 30–60 years in 1947. It has two advantages over other studies conducted at the same time: it was made up of the general population, i.e. the population was not selected, which eliminated the bias associated with the selection of participants; and, unusually, it included 2,873 women, which would allow for later tracking of the onset of heart disease by sex.

Thomas Dawber, director of the study for 15 years, from 1949 to 1965, described the study design and its implementation mobilizing medical personnel from nearby hospitals. Using this test population, the objective of the study was to improve understanding of the development of cardiovascular disease and to detect the early signs of disease. The clinical examination record was taken every two years. Interpretation of the participants’ diagnostic and exposure data was standardized according to the recommendations of a technical committee of cardiologists. Through the rigor of its results, the Framingham survey has made it possible to identify the main risk factors for cardiovascular disease, such as high blood pressure, high blood cholesterol, smoking and obesity. It is still active today with its third generation of participants; it is a prospective cohort study model [SPL 07].

What samples and data are collected in this pioneering biobank?

The scientific committee defined the descriptive, medical and biological examinations for the participants [DAW 51]. Each individual was subjected to a thorough medical questionnaire about their medical, family and lifestyle history. Next, the personal examination recorded each participant's anthropometric parameters (weight, height, chest and waist circumference, eye and hair color, etc.), health status (presence of tumor, liver hypertrophy, femoral pulse, etc.) and, more specifically, cardiac constants (heart rate, blood pressure, etc.). The blood samples and clinical examinations dictated by the committee were attended to in nearby hospitals.

Biological constants were monitored and interpreted by the same committee based on:

– blood samples for biochemical (glucose, hemoglobin, blood cholesterol, serum phospholipids, uric acid) and serological (syphilis test) assays;

– urine samples (routine analysis);

– clinical examinations (X-ray and electrocardiogram).

Dawber et al. invented the *population biobank* by collecting patient information from medical history, physical examinations and laboratory tests. They associated the characteristics of each individual with the potential development of cardiovascular disease. They classified the population into two groups: the first with symptoms of arteriosclerosis or cardiovascular disease and the second without them, the healthy population.

1.1.3. *Classification of human sample biobanks*

In 2001, the OECD formalized the role of human sample collections, historically called BRCs. For the OECD, BRCs are "centers for the preservation of living cells, the genome of various organisms, and information on the heredity and function of biological systems. BRCs must meet the high standards of quality and expertise demanded by the international research community and industry for the dissemination of biological information and materials" [OEC 01].

The European Commission Joint Research Centre (EC-JRC) lists the five general missions of a biobank:

– collect and store medically and sometimes epidemiologically annotated biological samples;

– conceive of sample collections as dynamic – not static – projects that evolve over the long term;

– associate the specimens with current and future research projects;

– code donations to ensure confidentiality of data, while allowing the patients' identities to be traced if clinical information were to be transmitted to the patient;

– respect the protection of the rights of the donors and the biobank managers.

There are several types of biobanks that meet these criteria. A first distinction can be made according to the mode of financing, which distinguishes between public and private biobanks. Public biobanks have mainly been developed within university hospitals and cancer centers by benefiting from recurrent public funding. Private biobanks are set up by pharmaceutical companies and contract research organizations (clinical research organizations). Their activity is focused on clinical trials. Private funds can be invested in public biobanks, thus creating a mixed model of public–private partnership. The second distinction is the type of biobank, which is the most common designation. It concerns the donor group and the purpose of the biobank, depending on whether it is for the study of diseases that will eventually be developed in a population or the specific study of a single disease, such as cancer.

The three main groups are population biobanks, clinical biobanks and tumor biobanks, as developed below.

1.1.3.1. *The population biobank*

Population biobanks, such as national biobanks, collect samples from the population without selecting participants, with the goal of reflecting the general population. Typically, blood samples are collected at the beginning of the study, along with associated data (medical and family history, lifestyle, environmental exposures, etc.), and then the tests are followed up for several years. The strength of these studies lies firstly in the observation of the occurrence of the disease in a healthy population, secondly in the identification of genetic variants predisposing to the disease and finally in the identification of environmental risk factors. This prospective approach is essential in preventive medical programs. Thanks to population biobanks, it is possible to study predictive biomarkers when the participants have not yet developed the disease. The only drawback is that the study must be conducted over a long period of time (10–15 years), which is usually the time frame in which the disease develops.

The best example of a population-cohort is the Estonian genome center of the University of Tartu (EGCUT). The project was launched in 2018 with the goal of sequencing the genome of the entire Estonian population of 1.3 million [LEI 15]. Currently, the cohort size is 200,000 adult donors, reflecting the population distribution by age, gender and geographic location. The biobank has DNA, plasma and white blood cells for each donor. The aim of this biobank is to collect (genetic) information from the Estonian population and to use it as a preventive tool for personalized medicine. The originality of this biobank is that it is a public biobank for large-scale DNA collection and analysis.

1.1.3.2. *The clinical biobank*

Clinical biobanks dedicated to a pathology have a selection criterion. The specimens collected come from patients suffering from a particular disease. Collected during the patient's medical diagnosis or treatment, these pathological samples may be biological fluids (blood and urine) or tissues. They are often stored at the point of collection, i.e. in hospital biobanks, and stored for research purposes.

Integrated in the University Hospital of Nice, the Côte d'Azur biobank (Laboratory of Clinical and Experimental Pathology, LPCE06) is dedicated to lung cancer. Washetine et al. analyzed the organization of this biobank since its beginning in 2006 using a quantitative and qualitative approach [WAS 18].

Quantitatively, the collection currently includes nearly 4,000 biological specimens from thoracic surgeries performed at the University Hospital of Nice. The main epidemiological data are listed: histological type, molecular characteristics of lung adenocarcinoma and tumor stage. Five tumor tissue samples and five non-tumor tissue samples are frozen for the same specimen. Formalin-fixed paraffin-embedded (FFPE) blocks were also collected for the majority of patients (94%). Blood samples are obtained for 67% of patients.

Qualitative data is optimized by reducing ischemia and transport time to the laboratory. Thanks to the transport of the specimens by pneumatic tube, cold ischemia lasts about 30 minutes. Blood samples, taken before surgery, are prepared immediately through centrifugation, separation of circulating blood cells, sampling and freezing at –80°C for plasma and serum and in liquid nitrogen for blood cells in the presence of DMSO.

a)

b)

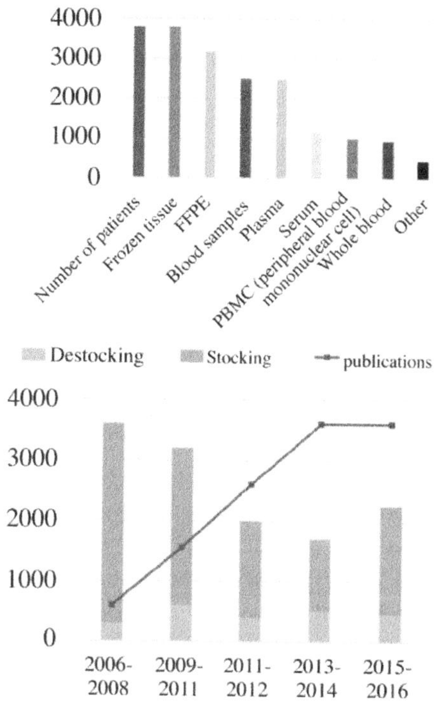

Figure 1.2. *Activity indicators of the Côte d'Azur biobank. a) Quantitative data for numbers of samples stored in the biobank. b) Storage and withdrawal of samples according to the given periods and number of citations mentioning the Côte d'Azur biobank. For a color version of this figure, see www.iste.co.uk/arrighi/biobanks.zip*

COMMENTARY ON FIGURE 1.2.– *The data are taken from the work of Washetine et al. with their kind permission [WAS 18].*

The analysis of the biobank's functioning has highlighted weaknesses such as the absence of specific collections (feces, saliva), the low number of stage III and IV tumors and the accumulation of samples without request for inclusion. One of the strategic actions to be implemented as a priority is gaining access to samples from clinical trials, after authorization from the local ethics committee, in order to systematically obtain samples from stage III/IV patients.

The standardized quality procedures enabled the NFS 96-900 to be obtained in 2010 and renewed in 2013 and 2016. The samples were collected by the ISO 15189 accredited hospital laboratory in 2013. These standards guarantee the biobank's partners high-quality samples that are perfectly annotated clinically.

1.1.3.3. *The tissue biobank or tumor biobank*

The French National Cancer Institute (*Institut National du Cancer* (INCa)) defines *tumor biobanks* as "infrastructures for the cryopreservation of tumor samples from cancer patients, in response to medical and health obligations and scientific objectives". In France, there are 58 tumor biobanks in the main university hospitals and cancer centers. Twelve of them are located in Paris within the *Assistance publique des hôpitaux de Paris*, which is the university hospital system operating in Paris, the largest hospital system in Europe.

Tumor libraries respond to the need for genomic classification of tumors and the search for molecular markers in cancerology. By providing samples and associated medical data, tumor libraries allow researchers to develop large-scale transcriptomic, proteomic or metabolomic analyses, making it possible to identify new prognostic or diagnostic *biomarkers and new therapeutic targets*. The identification of these biomarkers has also allowed the development of "*companion tests*" for the use of targeted therapies. The French National Authority for Health (*Haute Autorité de Santé* (HAS)) defines a companion test as "a diagnostic test that makes it possible to select, based on their status for a predictive biomarker identified by this test, only those patients for whom treatment is likely to be beneficial, among those diagnosed with a given disease. For this reason, the test is referred to as a 'companion' to the treatment".

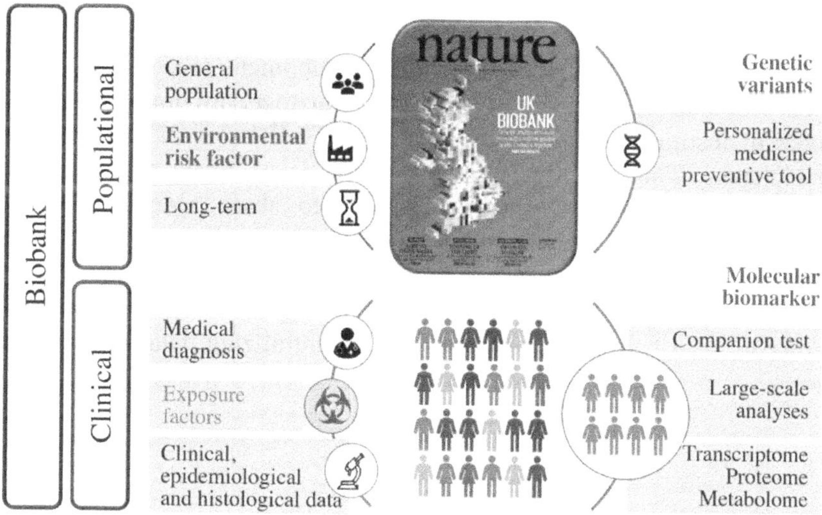

Figure 1.3. *Population and clinical biobanks. For a color version of this figure, see www.iste.co.uk/arrighi/biobanks.zip*

COMMENTARY ON FIGURE 1.3.– *Depending on the purpose of the collection, biobanks are population-based if they keep specimens from the entire population for a long time. If they select pathological samples, they are clinical biobanks dedicated to a given pathology in order to understand the molecular and cellular mechanisms and discover therapeutic leads.*

The discovery of new candidate biomarkers leads to retrospective studies, enabled by large sample collections accompanied by epidemiological, clinical, histological and genomic data. These data can be shared in a national or international network of several biobanks. In the following section, we locate the major biobanks by presenting their objectives and the first results of large-scale cohorts.

1.2. Mapping of biobanks

The coordination center for European biobanks (the Biobanking and Biomolecular Resources Research Infrastructure (BBMRI)) is located in Graz, Austria. Ten years ago, Professor Kurt Zatloukal, a pathologist at the Medical University of Graz, had the idea to connect European biobanks. He was the coordinator of the BBMRI project, funded by the European Strategy Forum on Research Infrastructures (ESFI) under the 7th European Framework Program for Research and Development (FP7). This project started with a dozen countries in 2008 and today there are more than 30. The number of institutions that have joined the BBMRI project reaches nearly 280 biobanks. The objective of this network is to create a platform that connects researchers and allows scientists to share samples and their associated data. In a first step, the goal of this infrastructure was to list the biobanks in Europe and to methodically identify all the collections. Then, in a second step, identical and standardized operating procedures were imposed in order to meet harmonized routines when centralizing data in this unique network.

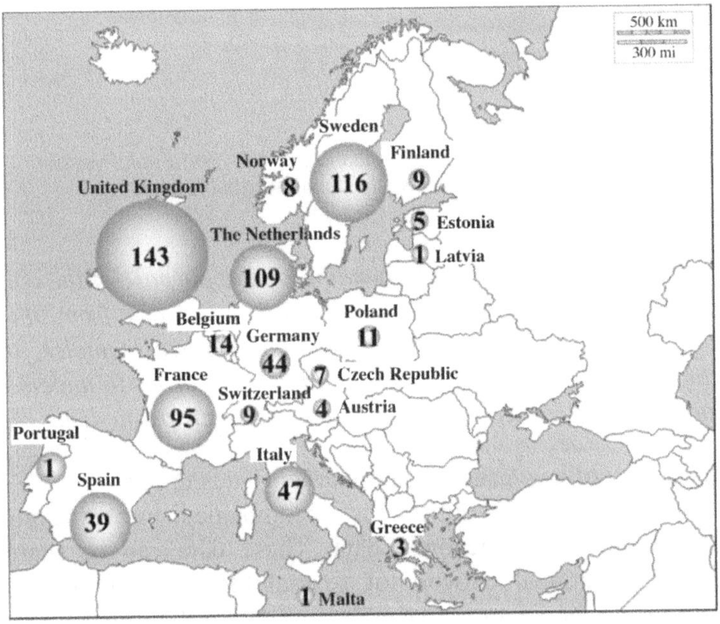

Figure 1.4. *Map of biobanks in Europe. The number of biobank members of the BBMRI-ERIC infrastructure is indicated by country*[1]

Between 2014 and 2020, the research and innovation program of the European Union Horizon 2020 planned to provide Europe with world-class research infrastructures. It is in this context that *the research infrastructure*

1 Portal data: www.bbmri-eric.eu/, May 2021.

for biobanking, named the Biobanking and Biomolecular Resources Research Infrastructure–European Resource Infrastructure Consortium (BBMRI-ERIC), has been funded. The objective of this infrastructure is to bring together all biobanking stakeholders – researchers, biobankers, industry and patients – to accelerate medical research and the discovery of new treatments.

It is the largest European infrastructure dedicated to human health research. It helps researchers find the sample and data they need for their research project while respecting the quality and regulatory constraints imposed on biobanks. BBMRI-ERIC also facilitates interdisciplinary collaboration (informatics, legal and ethical).

1.2.1. *The large catalog of European biobanks*

The first version of the BBMRI-ERIC catalog, released in July 2015, consisted of a directory of information on biobanks. A few months later, in December 2015, this information tool was enriched with additional data on biobank collections and networks. In practice, the BBMRI-ERIC catalog is used to locate samples or services that are essential to the advancement of research, thus identifying biobanks that hold these samples or are able to offer the expected services.

Note that this is not personal data. All data are public and can be shared freely. Thanks to this interface, all biobank actors, as well as researchers and patients, can access the general data for each biobank:

– *biobank users* publish information on the site about a biobank's activity and gain visibility. Informed of the composition of the collections and their location, they can initiate collaborations with these biobanks;

– *researchers* identify the samples and associated data and contact the biobanks that hold the collections they need to conduct their research project;

– *patients and donors*, made aware of the needs of research through communication campaigns, can participate by giving a biological sample to the biobank of their choice.

This database describes the samples and general data of each biobank; it is an information tool between biobanks. Contrary to existing systems, the architecture of this tool centralizes an online and offline information flow with a possible connection to the biobank management system. This interoperability makes the database sustainable while providing an interactive interface and automatic updates.

The web application is built on the MABIS 2.0 (minimum information about biobank data sharing) standard, an open source software adopted by biobanks and research communities. The work of van de Velde et al. has shown that this bioinformatics data software is aimed at non-bioinformaticians. It collects, manages, analyzes, visualizes and helps to share extensive and complex biomedical data without requiring advanced bioinformatics skills [VAN 19].

The development of the database was defined to integrate three features:

– storage capacity for biological materials and derived products;

– data storage capacity;

– the integration of the expertise of biobank users.

The work of Holub et al. detailed the contents of this massive catalog of European biobanks [HOL 16]. In 2016, the authors identified 515 biobanks and estimated the total number of samples in 136 clinical and 189 population biobanks in Europe to be over 60 million. In 2020, the BBMRI-ERIC brought together 19 member countries of the European community and an international organization, the International Agency for Research on Cancer (IARC). They cover a population of more than 500 million individuals in Europe. This large catalog of European biobanks inventories 100 million samples [BBM 21]. In addition to the ability to select data by country, by biobank name or by pathology, other selection criteria facilitate the search for resources – sample or biobank – such as the nature of the sample, the type of collection, the type of data and the quality labels of the biobank and the collection.

Thus, the biobank repository provides the researcher with access to cohorts of individuals with a specific phenotype and guarantees that these biological samples have been collected, processed and stored according to standard quality procedures.

Material		Collection	Data
Blood	Whole blood		
	Blood cells	Healthy controls	
	Plasma	Birth cohort	
	Serum	Twins	Biological sample
Tissue	Freezing	Cross-sectional study	Genetic investigation
	Paraffin blocks	Longitudinal study	Imaging data
Nucleic acids	DNA	Population	Medical records
	cDNA or mRNA	Hospital	National registries
	RNA	Specific disease	Physiological and biochemical analyses
	microRNA	Rare disease	Surveys and studies
Cell line		Image collection	
Saliva		Non-human	
Urine			
Feces			
Quality: certification of the biobank/collection by an accreditation body and/or BBMRI-ERIC audit of the biobank and the collection			

Table 1.1. *Criteria for locating a collection and identifying a biobank in the BBMRI-ERIC database*

COMMENTARY ON TABLE 1.1.– *Selection is made via a drop-down menu. By choosing a blood material, for example, the sub-menu proposes several derived products, such as circulating blood cells, plasma or serum*[2].

1.2.2. *International biobanks*

As soon as the samples (blood, tissue, urine, etc.) arrive at the biobank, they are received and the donor's personal information, such as their name, surname, age, gender and primary diagnosis, is entered into the database. Then, the sample is identified with a barcode linking it to the patient's

2 Portal data: https://directory.bbmri-eric.eu/menu/main/app-molgenis-app-biobank-explorer/biobankexplorer, March 2020.

medical record, and the sample undergoes a processing step before storage. It is divided into several fractions or transformed into derived products, such as nucleic acids, for example, after an extraction protocol. All products derived from the initial specimen are stored and made available to researchers.

The sum of the numbers of specimens (or fraction of specimens) and derivatives is an indicator of the total amount of samples stored and thus of the size of a biobank. However, there is a second indicator previously estimated during a collection project, the target size of the biobank. It aims at a maximum number of participants, knowing that a participant can provide several samples and that a sample can generate several derived products. These two indicators are publicly available because they indicate the order of magnitude of the size of the biobanks. According to the information they publish on their websites, the three largest biobanks in the world are located in Europe, China and the United States. While the size of the biobank and the number of samples are important, the purpose of the biobank is crucial. Is it a public, not-for-profit organization? Is it a population biobank? All of these questions need to be asked in order to understand how these infrastructures work.

1.2.2.1. *The Graz Biobank as a hub for European biomedical research*

Founded in 2007 and located in Austria at the Graz University Hospital (Landeskrankenhaus-Universitätsklinikum Graz), this biobank was conceived as the research center of the Medical Faculty of Graz. This public non-profit organization is partly financed by the medical faculty for the expenses related to staff salaries and overheads. The development of the infrastructure is supported by the Austrian government, the city of Graz and the BBMRI network through the European Union's Horizon 2020 research and innovation program.

Since 2009, the biobank has been ISO 9001:2008 certified; it has developed a rigorous quality management system based on standard operating procedures. It has undertaken pioneering work in the automation of sample storage for two reasons: firstly, to reduce transfer times during retrieval and ensure optimal sample quality, and secondly, to adopt a professional organization and reduce errors during sample retrieval. All ultra-low-temperature freezers (–80°C) used for the storage of biological fluids such as serum, plasma, blood cells and nucleic acids, as well as all liquid nitrogen tanks used for the storage of tissues and certain cell lines (–196°C) are under continuous monitoring. Each piece of equipment is equipped with a temperature recorder with an alarm signal in case of a power failure (freezers' power supply stop) or a decrease in the volume of liquid nitrogen in the tanks (release of nitrogen gas from the container).

Currently, the Graz Biobank has combined the collections from the Graz Medical School and the Institute of Pathology. These FFPE samples have been stored since 1984 – thus for more than 30 years – and represent more than 1 million patients and more than 20 million samples of various types, such as paraffin-embedded (FFPE) or cryopreserved tissues and biological fluids (whole blood, plasma, serum, urine, cerebrospinal, follicular or seminal fluid, etc.), and all these samples are strictly linked to the patient clinical information [BIO 21]. The Graz Biobank has done a great deal of work to ensure that the patient medical and personal information is accessible and confidential. Every sample is linked to the clinical information in the medical record, diagnostic and treatment results, as well as histological, medical imaging, immunohistochemical and molecular genetic characterization data.

The Graz Biobank is a clinical biobank in the broadest sense. Any information from a patient registered in the clinical database of the Graz University Hospital can be linked to the sample deposited in the biobank provided that the patient is of age and able to give consent (no samples from psychiatric patients). All patients and donors sign a free and informed consent form that authorizes the use of their samples for medical research in the form of extended consent. Biospecimens are routinely collected at diagnosis or during therapeutic interventions, which are performed on an annual population of about 400,000 outpatients and 85,000 inpatients, according to data from Huppertz et al. [HUP 16].

Since 2012, all samples are coded in the form of a QR code (2D datamatrix code) that aggregates the minimum patient information: identification number, age, gender and primary diagnosis. In addition, detailed information is entered into the hospital's clinical information system in connection with the sample identification.

During clinical diagnosis, all results from the Graz Medical University's Institute of Pathology are made available to the biobank for fresh frozen and paraffin-embedded samples. In the case of tumor tissue sampling, a sample of normal tissue close to the tumor is taken together with the affected tissue. This serves as a non-tumor control tissue (fresh frozen or FFPE). In addition, healthy donor samples are collected in parallel and serve as control samples of serum, EDTA plasma and buffy coat (white blood cells and platelets). As the samples are collected by the hospital during surgery without prior selection, this collection reflects the Austrian population without any filter. In addition, the biobank also has targeted collections, such as cohorts for cardiovascular diseases, fertility disorders, febrile diseases and osteoporosis.

Huppertz et al. examined the activities of the Graz Biobank using both a quantitative approach, involving the increase in the volume of samples collected over the last 30 years, and a qualitative approach, involving the automation of sample labeling and tracking [HUP 16]. They see this infrastructure, which is in permanent interaction with hospital services and scientific research, as the hub of European biomedical research.

1.2.2.2. *China National Genebank (CNGB)*

The first Chinese biobank was created in 1994 when Chu's team discovered cell lines specific to Chinese ethnic groups [CHU 98]. Under the impulse of the Chinese Academy of Sciences, which was anxious to preserve these precious cell lines, the program for the creation of human biobanks was started in China. The Chinese government and the academy perceived the importance of the conservation of biological resources, especially human resources. They have massively funded precision medicine research projects requiring the construction of huge biobanks backed by biomedical research. According to the history of the study provided by Gan et al., the number of new biobanks has exploded in the last 10 years [GAN 15]. Clinical biobanks have been set up in large cities such as Beijing, Shanghai and Guangzhou near medical research centers and large hospitals.

More recently, China's first national genebank, inaugurated in 2016, has been described as one of the largest and most modern. Named the China National Genebank (CNGB), it is a public non-profit organization located in Shenzhen in southern China [CHI 21]. It consists of a bioresource bank, a bioinformatics data network and an operational system deployed in five interactive units, including three banks and two platforms:

– a biorepository where more than *10 million samples* of human, animal, plant and microbial origin are stored;

– the bioinformatics center for the storage and interpretation of transomics data (genomics, transcriptomics, proteomics and metabolomics);

– the "living" bank intended to cryopreserve living biological material, and more specifically to safeguard more than 300,000 plant species and millions of animals and microbes;

– the digital sequencing platform;

– the synthetic biology and genome editing platform.

The CNGB's mission is to collect and store genetic and biological samples from human beings and other living organisms. Its major activities focus on sequencing and genetic engineering applied to health, agriculture, microbiology and marine biology. The CNGB works in partnership with about 100 international organizations, including biobanks, for example, the Svalbard global seed vault in Norway, and international networks, such as the International Society for Biological and Environmental Repositories (ISBER).

The CNGB is part of the Beijing Genomics Institute (BGI), the number one genetic sequencing company in China and one of the world's leading genomic companies, along with the Sanger Institute in the UK and the Broad Institute in the US. The BGI group is a complex semi-private structure made up of various branches:

– public non-profit organizations such as the China National Genebank, the Gene Research and Training Center (BGI Research, BGI College) and Gigascience Scientific Publishing;

– subsidiaries dedicated to genetic sequencing (BGI Genomics) and agriculture (BGI Agro).

Created in 1999 in Beijing by the Chinese Academy of Sciences, the BGI set up its main office in Shenzhen and began its activities with participation in the human genome project. According to David Cyranoski's analysis, Chinese life science research is mainly oriented towards massive sequencing. In the scientific journal *Nature*, Cyranoski analyzes the role of the BGI, which could become "the world leader in genome sequencing" [CYR 10]. According to David Cyranoski, the BGI is becoming a "genomics factory" equipped with hundreds of high-throughput sequencers that could theoretically sequence 10,000 genomes per year. The Chinese group has

4,000 scientists and technicians, including 1,500 bioinformaticians, according to the firm's 2011 data.

The BGI has published its sequencing work in the journals *Nature* and *Science*, decoding the genomes of several plant (rice [YU 02] and cucumber [HUA 09]) and animal species (silkworm [XIA 04], chicken [INT 05] and giant panda [LI 10]), as well as the human genome (for example, in a prehistoric human preserved in the permafrost of Greenland for 4,000 years [RAS 10]).

In 2003, the BGI sequenced the genome of the coronavirus [QIN 03] responsible for severe acute respiratory syndrome (SARS). In 2019, a new species of coronavirus appeared, causing a viral pneumonia first localized in Wuhan, China, which then spread to all continents, triggering a global pandemic. Its genome, sequenced by the BGI, confirms that it is a virus previously unknown in humans, officially named Covid-19.

The Covid-19 virus genome was sequenced from bronchoalveolar lavage fluid samples from nine hospitalized patients in Wuhan. It is close – with 88% identical sequences – to the bat-hosted SARS coronavirus found in 2018 in the Zhoustan region of eastern China. It is more distant from the human coronavirus SARS (79% identical sequences) and MERS (50% identical sequences) [LU 20]. Lu et al.'s phylogenetic analysis suggests that bats are the original hosts of the virus, which was transmitted to intermediate hosts, animals sold in the Wuhan market, before crossing the species barrier and infecting humans. The molecular structure of the virus has shown that it binds to the human angiotensin-converting enzyme 2 receptor, which is the preliminary and essential step of viral amplification.

Based on a *reverse transcriptase PCR* (RT-PCR) approach, the BGI developed a molecular detection test for Covid-19. The primers designated and synthesized by the BGI were tested on the nine patients in the study. The results were then validated with a different protocol by the China disease control and prevention center (CDC).

This real-time fluorescent RT-PCR kit for detecting SARS-2019-nCoV includes the reagents and controls necessary to test approximately 40 patients in a few hours. It has been approved by the Chinese authorities, certified in Europe, and the US Food and Drug Administration (FDA) has approved it for emergency use in the US clinical market [BGI 21].

1.2.2.3. *The American genetic biobank 23andMe*

The French daily newspaper *Le Monde* profiled Anne Wojcicki, who runs the company 23andMe, in the article *Anne Wojcicki et 23andMe, les tests génétiques à portée de clic*, which appeared in the *Talents du net* series on July 19, 2018. After studying biology at Yale University, Anne Wojcicki worked for a decade as a financial analyst in the biotech industry. In 2006, she and Linda Avey, a biologist like herself, founded 23andMe, the leading Californian company in the genetic testing market, "23" being the number of pairs of chromosomes in humans.

Through the Internet, 23andMe goes directly to the consumer and markets a genetic test related to health and genealogy. It is available over the counter on the Internet without a medical prescription. This test can provide genetic risk factors for diseases such as cancer. In 2008, *Time* magazine awarded the Invention of the Year prize to the saliva-based genetic test developed by 23andMe. Anne Wojcicki challenges the concept of health industries; she wants to give consumers access to personal health information by putting genetic tests in their hands. However, without any guidance or interpretation of the results by a doctor, 23andMe's offer remains uninformative for the consumer.

At times, access to genotyping has led to irreversible consequences, such as self-medication treatment changes without the advice of a physician, or unnecessary surgeries, such as mastectomy, based on a false-positive result. In 2013, the FDA questioned the quality of health-related genetic testing kits and called for proof of the reliability of results and explanation of results to consumers. It banned the tests from marketing pending compliance with regulations on predictive tests for medical risk factors, subject to medical device marketing and quality rules. From 2013 to 2015, the FDA exercised regulatory control over health tests, and only ancestry-related tests were allowed. Internal company changes resulted in FDA approval for marketing towards approximately 30 autosomal recessive diseases, such as cystic fibrosis, sickle cell disease and Tay–Sachs disease.

Along with MyHeritage, LabCorp and Myriad Genetics, 23andMe is one of the major players in the global DNA testing market. Stoeklé et al. analyzed the *new business model* of these biotech companies, whose specialty is to produce, process and analyze genetic data on a large scale. Stoeklé's study divides their business model into two segments and describes 23andMe's model as a "two-sided data-banking market model" [STO 16].

This business model, which emerged with the explosion of the digital industry, manages data in two categories. On the one side, the company offers free services to the consumer and stores their data. On the other side, the consumer's data is sold to other companies. The content of the data differs from one company to another, but the example of Facebook illustrates the model of sharing personal information between users and also between companies.

In the case of 23andMe, the company collects biological samples through the offering of a direct sales genetic test at an attractive price. The protocol involves three simple steps (order–spit–discover). The customer orders a kit – health or genealogy – by creating an account on the Internet (order), where they are asked to fill out a form where information about their health, family, environment and lifestyle is declared. A few days later, they receive the kit containing a sampling tube labeled with a personal barcode into which they deposit a few milliliters of saliva (spit). The saliva sample is returned to the company by prepaid mail, analyzed and the results are sent via the Internet to a personal account within three to five weeks (discover).

The company 23andMe stores in its research division:

– biological samples;

– sequencing data;

– medical, family, personal and lifestyle information.

This is the definition of a biobank: 23andMe is one of the first biobanks to have integrated web tools. One of its first investors was Google, and the digital giants are now looking to invest in the market for genetic, genealogical and more broadly health data.

When announcing the collaboration with GSK in 2018, Anne Wojcicki explained that "I started 23andMe with the belief that by getting people interested in learning about themselves and participating in research, we would all benefit" in a post appearing on the 23andMe blog on July 25, 2018 [WOJ 18]. Anne Wojcicki holds the principle of sharing everyone's data for the health of all, which is what led to the sequencing of the human genome (Personal Genome Project (PGP)). This idea of sharing is implemented by free access to PGP data (open access), but access to 23andMe data is forbidden, locked by the company. Different from "sharing", Stoeklé sees it as a "commercial exchange", giving biological data a market value. He recognizes one side of this business model as the collection of samples, where, for a relatively low introductory price, consumers gain access to their genealogical information and health risks. The second side compiles genetic data and personal information and allows clinical annotations to be added to the biobank samples. This makes them attractive to pharmaceutical companies who covet large-scale cohorts to detect disease exposure factors.

In France, open access genetic testing is prohibited. According to article 16-10 of the Civil Code, paragraph 1: "The examination of a person's genetic characteristics may only be undertaken for medical or scientific research purposes." However, many French people buy these kits online and enrich the French genetic data biobank. Companies have collected biological samples linked to personal data declared online such as family, medical and lifestyle information. Eighty-five percent of customers have agreed to share their anonymized results with biotechnology companies (Genentech), large pharmaceutical firms (Pfizer) or multinationals (Procter & Gamble).

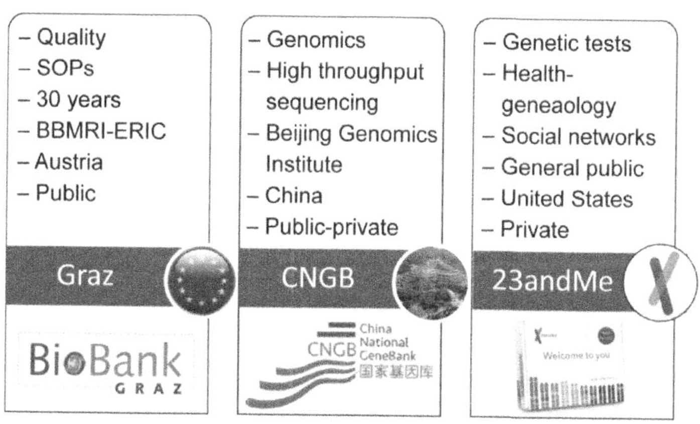

a) b) c)

Figure 1.5. *Strengths of large international biobanks. SOPs: standard operating procedures; BBMRI-ERIC: European biobank research infrastructure; BGI: Beijing Genomics Institute; NIH: National Institutes of Health*

COMMENTARY ON FIGURE 1.5.– *a) The Graz Biobank, the oldest, is known for its quality standards; b) the China National Genebank (CNGB) has enormous high-throughput sequencing capabilities and collaborates with the Beijing Genomics Institute; c) the private biobank 23andMe markets genetic test kits via the Internet and has more than 1 million clients worldwide participating in collaborative studies with several pharmaceutical companies.*

1.2.3. *The first results from the megacohorts*

Epidemiological cohorts are designed to study and follow up populations over long periods of time in order to monitor the development of various diseases. The goal is to identify genetic or environmental factors that lead to the development or progression of disease. Multifactorial studies are long term (10–15 years) and require the inclusion of hundreds of thousands of participants to detect the onset of disease. Two examples of "megacohorts" are presented: the French Constances cohort, which has been followed for nearly 20 years, and the EPIC cohort, which has been in place in Europe for about 30 years. "The data from the cohort allow us to analyze the health status of the population and to better understand what happens throughout life. After several years of recruitment, we have been able to analyze the data and provide initial estimates, in particular on the prevalence of being overweight and obesity", explains Marie Zins, epidemiologist and scientific and technical manager of the Constances cohort.

1.2.3.1. *Mapping obesity in France according to the Constances study*

"One out of two French people is overweight" was the headline of Pierre Le Hir's article in the newspaper *Le Monde* on October 25, 2016 [LEH 16]. It paints a portrait of the French man – and the French woman – based on the first results of the Constances cohort, published by Matta et al. in the weekly epidemiological bulletin of Santé publique France [MAT 16]. What is the Constances cohort? It is a major epidemiology and public health research project in France, which was launched in 2012. The cohort now has more than *200,000 participants*, and more people continue to join regularly. Conducted by Inserm and the *Caisse Nationale d'Assurance Maladie des Travailleurs Salariés* (CNAMTS), this longitudinal study tracks the health of French people, their diet, their environment and their working conditions. It brings together the scientific forces of several countries; about 80 research projects are underway and 11 French and international research consortia are participating. The number of scientific publications is accelerating, from three articles in 2017, to 18 in 2018, then 28 in 2019, 23 in 2020 and 25 in December 2021, according to the PubMed NCBI bibliographic database.

Every year, participants complete a health questionnaire, and every five years they undergo a medical examination at a health insurance center. Since 2018, sample collection has been organized in partnership with the *Integrated Biobank of Luxembourg* (IBBL). It provides the necessary kits for the collection, and then takes care of the transport, processing and storage of the samples, guaranteeing their quality and traceability. Ninety-five percent of the participants have given their consent, and 250,000 samples – plasma, serum, buffy coat and urine – from 9,500 volunteers are stored at the IBBL. These samples are available to researchers who can request them by submitting a scientific project to Constances, which will be validated by the International Scientific Council and the Industrial Steering Committee. As for any project involving human samples, authorizations from the Institutional Review Board (*Comité de Protection des Personnes* (CPP)) and the *Commission Nationale Informatique et Libertés* (CNIL) are regulatory and essential.

"The scientific interest of the Constances biobank is to redistribute all these samples to researchers. These samples will help to elucidate the mechanisms of complex diseases, to understand the role of interactions between genetic, epigenetic and environmental factors in the development of diseases and to discover new biomarkers", explains Marie Zins, coordinator for Inserm of the Constances cohort.

The prevalence of being overweight and obese was studied in 29,000 participants aged 30–69 years, with as many men as women distributed in 16 French departments [MAT 16]. Two indicators were collected simultaneously: body mass index (BMI) and waist circumference. BMI is calculated from weight and height (weight in kg divided by the square of height in m), and usual marker of obesity is if it exceeds 30 kg/m^2, and of overweight above 25 kg/m^2. The results show that *one French* person *in two is overweight*: 56% of men and 40% of women are overweight, combining excess weight and obesity. Similarly, abdominal obesity, measured by waist circumference exceeding 94 cm for men and 80 cm for women, affects 41% of men and 48% of women.

Matta et al. showed a higher prevalence with age and a strong correlation with income level. Regional disparities were mapped by department: Paris was the least affected department (10%), whereas Nord (25%) and Meurthe-et-Moselle (22%) had high prevalences. At the individual level, the study found that 30% of women with low monthly income (less than 450 euros) were obese, compared to 7% of women with a much higher income (more than 4,200 euros).

The results of Constances concern many research themes around aging, social and professional health determinants, women's health, genetic factors and interactions with the environment. One of the themes, stress at work, has demonstrated the chronic use of benzodiazepines, anxiolytic drugs with a high risk of dependence. Using the pharmacy-dispensed treatments to more than 9,000 participants in 2015, Airagnes et al. calculated prevalences of benzodiazepine use of 2.8% in men and 3.8% in women [AIR 19]. Chronic benzodiazepine use is thus particularly common in the general French population, and work stress is associated with an increased risk of chronic benzodiazepine use.

1.2.3.2. *The European Prospective Investigation into Cancer and Nutrition (EPIC)*

EPIC is a large European prospective investigation whose aim is to study the links between diet, lifestyle and cancer incidence.

The EPIC cohort was initiated by the World Health Organization (WHO) and launched under the coordination of the Lyon-based International Agency for Research on Cancer (IARC) with the support of the European Community's "Europe Against Cancer" program. Recruitment of participants took place from 1993 to 1999, involving more than *520,000 participants aged 35–70 years living in 10 European countries* (Denmark, France, Germany, Greece, Italy, the Netherlands, Norway, Spain, Sweden and the United Kingdom). The participants (153,427 men and 367,903 women) completed questionnaires about their diet, physical activity, alcohol and tobacco consumption, family history and lifestyle. They also gave blood samples, which constitute the International Agency for Research on Cancer (IARC) biobank, also called the IBB. For each participant, samples – plasma, serum, white blood cells and red blood cells – were collected. Each sample was aliquoted into plastic straws and divided into two batches, one stored locally and the other sent to the IARC biobank.

Initial results from the EPIC study show the role of fruits and vegetables in preventing digestive and respiratory cancers. The study by Bradbury et al. showed that the regular consumption of fruits and vegetables is associated with a decreased risk of developing cancers of the mouth, larynx, pharynx, esophagus, colorectum and lung [BRA 14]. The other cancers studied (stomach, biliary tract, pancreas, cervix, endometrium, prostate, kidney, bladder and lymphoma) did not show an association. These results reinforce the idea of cancer prevention through a balanced diet containing a daily intake of fruits, vegetables and whole grains. "This, however, is only likely to lead to significant cancer prevention in France if the nutritional message is accompanied by a general recommendation for lifestyle modification: not smoking, limiting alcohol consumption, increasing physical activity and avoiding obesity", comments Professor Elio Riboli, coordinator of the EPIC study and Director of the School of Public Health at Imperial College London.

From epidemic to prevention is the pathway traced by this study. The work of Trichopoulou et al. examined the effect of the *Mediterranean diet* on health and longevity [TRI 14]. By substituting olive oil (not widely consumed in northern European countries) for other unsaturated fats, they demonstrate that the "modified" Mediterranean diet is associated with increased longevity in older individuals. In all, more than 100,000 participants in the EPIC cohort over the age of 60 were recruited. Of the total, some participants were excluded, either because they had cardiovascular disease (15,362 participants) or because detailed dietary data were missing (10,340 participants), bringing the final number to 74,607 participants across nine European countries (France, Italy, Spain, the United Kingdom, the Netherlands, Greece, Germany, Sweden and Denmark). This observational study is a public health indicator in an increasingly elderly Europe.

The French part of the European EPIC study is represented by the epidemiological study of women in the French education system, called the *E3N cohort. The participants are female teachers* affiliated with the *Mutuelle Générale de l'Education Nationale* (MGEN), which supports the project alongside the *Ligue contre le Cancer*. The cohort includes approximately 100,000 women aged 40–65 years. They described their lifestyle and health status (cancer, cardiovascular disease, diabetes, depression, fracture, asthma, among others) by answering a questionnaire every two or three years. Biological samples supplemented these investigations with blood samples (whole blood, serum, plasma, erythrocytes and lymphocytes isolated in the buffy coat) and salivary and breast cancer tumor tissues.

The E3N cohort has produced broad results considering the complex role of diet, hormonal factors, physical activity and lifestyle in relation to health. According to Francoise Clavel-Chapelon, the major results concern hormonal treatments, being overweight and eating habits [CLA 15]:

– the benefit–risk balance of hormone replacement therapy during menopause, which has profoundly changed prescription habits in France;

– the association between chronic disease risk and anatomical characteristics;

– the influence of dietary habits on the risk of developing cancer, diabetes and asthma in middle-aged women.

Since 1990, the E3N cohort has been coordinated by Giluca Sevri and Marie-Christine Boutron-Ruault, both Inserm researchers. As an extension of this study, the *E3N–E4N family cohort* is composed of three generations: the women of the E3N study and their spouses (18,000 volunteers), their children and their grandchildren, whose recruitment began in 2021. The E3N–E4N study is interested in the influence of contemporary lifestyle on health in people of the same family over three generations.

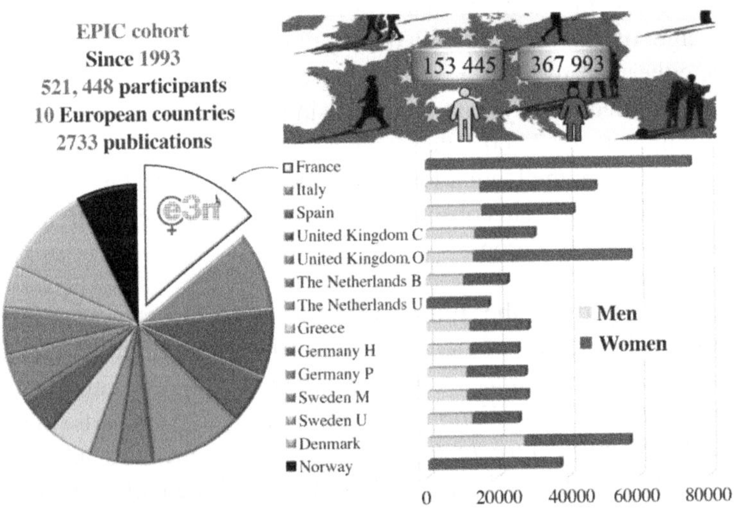

Figure 1.6. *The European Prospective on Cancer and Nutrition (EPIC). For a color version of this figure, see www.iste.co.uk/arrighi/biobanks.zip. UK C: Cambridge; UK O: Oxford; Holland B: Bilthoven; Holland U: Utrecht; Germany H: Heidelberg; Germany P: Potsdam; Sweden M: Malmo; Sweden U: Umea*

COMMENTARY ON FIGURE 1.6.– *Sometimes, two sites per country collected information and samples from the 153,445 male and 367,993 female volunteers. Like Norway and the Netherlands (Utrecht), France is represented by women only; this sub-cohort is named EPIC-E3N. The number of publications with the terms "EPIC" and "cancer" is given according to PubMed in December 2021*[3].

1.2.3.3. *The American precision medicine program: All of Us*

The *All of Us* research program is funded by the US government and led by the National Institutes of Health (NIH). Conceived under President Obama in 2015, the project was launched in 2018 along with the *precision medicine* initiative (PMI). Its goal is to recruit 1 million people in the United States to accelerate biomedical research and improve health. The program's challenges revolve around four major areas:

– identifying risk factors for certain diseases and discovering new biomarkers;

– understanding why treatments are effective in some people but fail in others;

– including all profiles in clinical studies to meet the needs of all;

– integrating new technologies into tomorrow's healthcare.

In short, this cohort will be a huge database for all prospective, retrospective and cross-sectional studies.

The recruitment of the *All of Us* cohort is the cornerstone of the project, hence the priority of raising awareness of the concept of medicine adapted to each person and convincing participants to enroll in this research program. In order to reach as many people as possible, a communication campaign has been launched on social networks and in the field, with a huge bus criss-crossing the country to meet the volunteers. The goal is to recruit 1 million adult participants by 2024 and to sequence 1 *million genomes*. Children will be included in a second phase. For 15 years, researchers will take dozens of blood and urine samples from each participant. At the same time, their health data, lifestyle and environment will be regularly monitored.

3 The information is taken from www.e3n.fr/epic.

The *Mayo Clinic* Biobank is responsible for the development of standards and protocols for the collection of specimens that will be stored and made available to researchers. Recognized as one of the world's leading health centers, the Mayo Clinic is a US research and teaching hospital located in Rochester, Minnesota. Its Center for Personalized Medicine is dedicated to promoting translational research and leveraging genomic data [LAZ 14].

The terms "personalized medicine", "individualized medicine" and "precision medicine" sometimes overlap, as the concept of personalized medicine remains unclear, as Bateman explains [BAT 14]. The idea of precision medicine entails a comprehensive view of the molecular and metabolic biology data, known as omics, that can characterize a individual:

– genomics: the sequencing of the individual's DNA (genome);

– transcriptomics: the set of transcribed messenger RNAs;

– proteomics: analysis of sample proteins;

– metabolomics: the set of small molecules (metabolites) present in the sample.

These data partially characterize the individual, as two other factors, intrinsic and extrinsic to the individual, are missing:

– the microbiome: the set of microorganisms present in the body;

– the exposome: the set of relations between the individual and the environment.

Finally, this complete profile lists all the factors that define the original and unique signature of each individual. This complex dataset is a haven of study for precision medicine to mimic, on the one hand, the individual, and, on the other hand, their belonging to a population and one of its subgroups.

Precision medicine needs a large number of participants who mirror the population. Very often, the scope of genetic studies is limited by small sample sizes, lack of diversity among participants or partial access to data, and thus the inability to link genotype and phenotype. Population-based genetic research requires very fine targeting of phenotypic data, which translates into access to participants of varied ancestry. The uniqueness of the *All of Us* cohort is its inclusion of minorities underrepresented in medical research, ethnicities and communities. The goal is to create the most diverse biobank in the world.

This program of sequencing 1 million genomes is only of value if the associated data are added to the genetic information. In this large-scale study, the project wants to develop the use of *new information technologies*, cell phones and smart biosensors to continuously monitor health status, physical activity and different health parameters. Participants will share their electronic medical records and answer questionnaires on their diet, sleep and environment, in short their lifestyle.

Ethical, legal and social issues (ELSIs) highlight the three objectives of the study:

– cohort diversity;

– participant commitment;

– data confidentiality and security [SAN 17].

Diversity and health disparities are expressed by socioeconomic status, age, geographic location, health status, literacy level and personal information technology skills. Inclusion of disadvantaged populations, people with intellectual disabilities and families with children is requested. Priority is given to ethnically or racially diverse populations, as genetic backgrounds may reveal genetic variants. Inclusion of indigenous nations such as American Indian and Alaska Native populations is reinforced.

Security and privacy are essential. However, there are several difficulties: firstly, the need to protect the patient – in particular their electronic medical record, genetic information and health data posted on the digital platform – and secondly, the need to open up access to the data to research organizations that would like to include them in their work. A third constraint is ensuring the patient's follow-up and providing individual feedback. This is not a simple task given the quantity and complexity of the information. As for the interpretation of the data, particularly the genetic mutations, this will be handled by Color Genomics, who will support the patient when they receive the results.

The *New England Journal of Medicine* Special Report provided an update on the progress of the All of Us program as of the end of 2019 [DEN 19]. After a testing phase in 2017, the recruitment process was validated, as well as data management and security. During this first phase, more than 27,000 participants were included. They committed to sharing their personal information, biospecimens and electronic medical records. After deploying 340 sites across the country, the recruitment rate was approximately 3,000 participants per week. The total cohort size was 230,000 participants. Of these, 51% were non-White and 80% were minorities who are typically underrepresented in medical research.

Access to the data will be available to the researchers from the beginning of the study and not after its closure, as is often the case. The data will be stored in a virtual space (Cloud) to which the researchers will have access through an access code, assigned once they have demonstrated the usefulness of these data for the advancement of their research.

Data access for participants is planned for the electronic medical record, routine biological analyses and genetic tests. It will be particularly monitored when results of genetic variants known to be risk factors for certain diseases are returned. They will then benefit from medical and therapeutic follow-ups.

In summary, this precision medicine program takes into account individual genomic differences, environmental factors and the American lifestyle. It creates a participatory health model, where the volunteers included in the All of Us cohort become the actors in looking after their own health. Francis Collins, Director of the NIH, sees a "coalition of partners" in "this largest and most ambitious health research project" on the scale of the American population [COL 16].

1.3. Process management in biobanks

Upon receipt of biospecimens in a biobank, patient consent is verified before the sample is registered. The flow of received samples is then monitored throughout the biobank.

The management of the whole process is divided into two parts: the management of the technical procedures that guarantee a high-quality sample and the management of the complex personal data that add value to the clinically annotated sample.

At each stage, the integrity of the sample is checked and maintained as it was when it arrived in the biobank. Maintaining the sample's high quality ensures high-quality results in biomedical research. *Technical factors* related to the sample such as transport (temperature, duration and mode of transport), processing (manual or automatic, duration and temperature) and storage (duration, temperature and freezing and thawing cycles) play an important role in the reproducibility of experiments. In the face of heterogeneous techniques, harmonization of operating protocols is required between the various methods and biobanks located around the world. Quality control and harmonization of procedures are the technical criteria for maintaining sample quality, accompanied by the automation of processing steps. These technical factors are essential for the collaboration of several research teams in a network.

In a human biospecimen biobank, *individual factors* are distinguished from technical factors. These individual factors depend on the clinical, genealogical, environmental and personal data related to the biospecimen donor. In medical research, these individual factors are as important as the others, so the biobank must be able to collect the personal and clinical data and then share it by clinically annotating each sample. The security of personal data and confidentiality of all information collected are the indicators of the trust between the physician, researcher and biobanker.

1.3.1. *Sample quality, the priority of biobanks*

Every year, the pharmaceutical industry spends billions of dollars on biomarker research in precision medicine. This research sets up collaborations between private and public partners to discover new biomarkers. The objective of partner biobanks is to rapidly build large-scale collections to accelerate the conduct of clinical and translational studies. Thousands of biological samples, such as tissues and blood samples, are cross-referenced with multiple analytical techniques such as immunohistology, genomics or, more broadly, omics. In order to ensure confidence in the analytical data generated by these various technologies, the quality of the biological materials and the associated data are the major challenges for biobanks. Throughout the sample's journey in the biobank, its quality is controlled, from reception (tissue or fluid), to preparation (extraction, centrifugation and aliquoting), then to conservation (storage at

–80°C or in nitrogen at –196°C) and finally release (transfer). Not only are biobanks confronted with the various types of samples, sometimes collected according to procedures that vary from one site to another, but they are also faced with the challenge of harmonizing the practices that they must put in place in order to guarantee research teams access to quality biological data and perfectly clinically annotated samples.

1.3.1.1. *Specific standards for biobanks: certification of professional quality*

In 2018, the International Standard Organization (ISO) published an international standard for biobanking, ISO 20387:2018 biotechnology – biobanking – general requirements for biobanking. Biobanks that have this label give users confidence in the overall functioning of the biobank and therefore in the samples transmitted. Moreover, this label increases the international visibility of the biobank. This ISO standard is defined by four points:

– it specifies general requirements for the competence, impartiality and consistent operation of biobanks, including quality control requirements, to ensure the appropriate quality of collections of biological materials and associated data;

– it applies to all organizations that conduct biobanking for research and development, including biobanking of biological materials from multicellular organisms (e.g. humans, animals, fungi and plants) and microorganisms;

– it recognizes and confirms biobanking activity with biobank users, regulatory authorities, peer review organizations and systems, accreditation bodies and others;

– it does not apply to biological materials intended for food/feed production or laboratories performing analyses for food/feed production and/or therapeutic use.

In France, the S96-900 standard for biobanks, implemented in 2008, combines quality management with technical skills in the field of biobanking.

To meet the criteria of these standards, biobanks can rely on a *quality management system* (QMS), which is based on the implementation of standardized procedures and audits. Thanks to a high level of standardization and standardized procedures shared between the different centers, the European BBMRI-ERIC network of biobanks has set up its own validation routines. This allows the creation of expert biobanking centers, in sync with the advances in pharmaceutical or medical research and analytical techniques.

1.3.1.2. *The impact of the pre-analytical phase on sample quality*

Faced with the lack of reproducibility of experimental results, estimated to result in costs of more than 56 billion dollars according to the study by Freedman et al., the challenge for biobanks is to guarantee the integrity of the samples and therefore to control the pre- and post-collection process up to storage in the biobank [FRE 15]. The pre-analytical phase begins as soon as the bloodstream no longer supplies blood to the sample. During a surgical procedure, the warm ischemia ends when the sample is removed from the body. The time it takes to get the sample to the laboratory extends the duration of this ischemia, which is then called "cold ischemia". The recommendations of the HAS suggest transport "without breaking the cold chain, in a cooled container saturated with nitrogen vapor or in dry ice". When the sample is received at the biobank, depending on whether it is solid or liquid, the conservation technique differs:

– a tissue sample is fixed in formalin and embedded in paraffin blocks at a controlled temperature of 55–58°C to avoid tissue degradation;

– a fluid sample is frozen in the presence of a cryoprotectant (glycerol or dimethylsulfoxide).

In addition to maintaining quality during transport, a second factor depending on the nature of the sample itself can affect the robustness of the results. The sample, of small volume or size, contains little material; however, it must be a representative of the sampling site (organ) and of the patient's physiopathological state (healthy or suffering from a pathology). If the sample contains a majority of necrotic cells and few healthy or tumorous cells, its analysis will not reflect the whole organ, healthy or affected. Hence, it is important to control *pre-analytical variables* such as the nature of the sample, the storage technique and its duration.

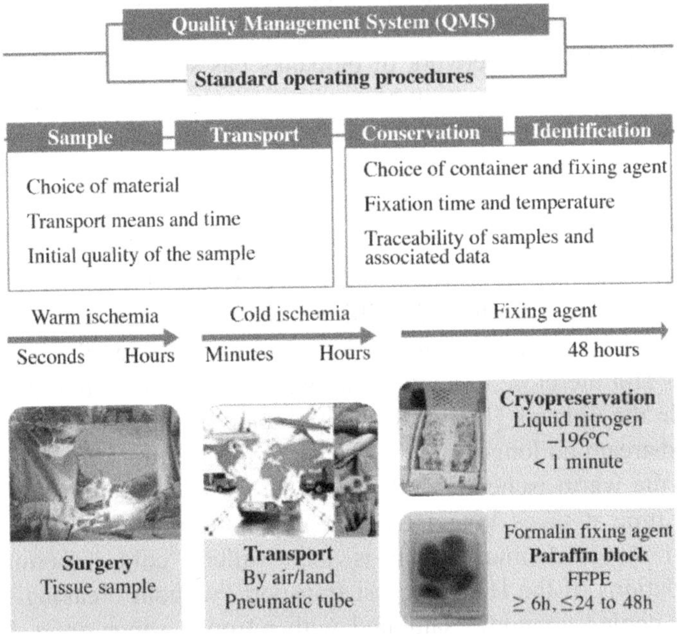

Figure 1.7. *Quality management system (QMS) activities. FFPE: formalin-fixed paraffin-embedded*

COMMENTARY ON FIGURE 1.7.– *Standard operating procedures are implemented at each step of the pre-analytical chain under the pathologist's supervision, starting with the surgical procedure, then the transport of the sample to the biobank and finally its storage. The data are taken from [AMA 20] with the authors' kind permission.*

FFPE tissues are stored at room temperature for decades without altering tissue integrity. Alternatively, after nucleic acid extraction, sample derivatives are stored at –80°C (freezer) or –196°C (liquid nitrogen container). At these very low temperatures, degradation enzymes such as ribonucleases are inactivated. The HAS recommend that these freezers be placed "in locked premises with air conditioning and effective alarm systems". These two preservation methods are the reference methods for genetic and histological analyses.

1.3.1.3. *Harmonized operating protocols*

Different types of processing and storage can alter the molecular composition of samples. To improve the maintenance of sample quality and consequently the quality of research resulting from these biological samples, it is important to accurately record the *pre-analytical history*. In a standardized way, the pre-analytical conditions are recorded in the form of a report (the pre-analytical coding for biospecimens (SPREC)) or more broadly in a biospecimen quality report (Biospecimen Reporting for Improved Study Quality (BRISQ)).

The BRISQ report is a data reporting tool that centralizes sample information. Biobank users enter all sample data and generate a report for researchers that can be included in publication appendices. The information describes the type of sample (tissue or fluid) and factors that may influence the integrity, quality and molecular composition of the biospecimen. These data do not guarantee quality or preclude quality control, but they do provide standardized information to researchers and regulatory agencies to improve the evaluation, interpretation and reproducibility of experimental results.

BRISQ recommendations are defined throughout the sample's journey through the biobank. Moore et al. constructed a hierarchy of information according to the four key stages of the process: sample preparation, collection, processing and storage [MOO 11]. The priority data consist of about 15 elements applicable to any type of sample, including the operative details of collection, stabilization and storage, as well as the clinical data associated with the patient. This first level of essential information is compulsorily documented in scientific publications. The second level includes additional data, i.e. 19 elements including the patient's personal information, stabilization times and temperatures. The last and third level of data are optional data that belong to the context and environment of the collection, such as parameters related to ischemia, therapy, health status, storage containers and transport. The biobank user must check the coded item and qualify each item with a short answer:

– biospecimen type;

– specimen's anatomical region;

– patient's health status;

– patient's clinical characteristics;

– collection method;

– type of stabilization after sampling;

– type of long-term storage of the sample;

– addition of a preservative;

– temperature and time for the transport and storage.

The first item defines the type of biospecimen, coded I.a., and is completed from the list of possible choices: solid tissue, whole blood, serum/plasma, isolated cells, urine, secretions or other human body derivatives.

The items in the BRISQ report are the result of discussions between researchers at the National Cancer Institute (NCI) and biobank users, including the International Society for Biological and Environmental Repositories (ISBER). After searching the biospecimen publications in the NCI bibliographic database, the BRISQ committee sorted the information according to its usefulness in the research project. The analysis of the information revealed three classes of information: parameters that were a priority to the research, those that were less of a priority and those that provided general information but did not advance the topic. Based on the terminology used in the bibliographic archive, the recommendations were prioritized, resulting in the BRISQ report. With this common terminology, information can be categorized in order of importance, including pre-analytical variables. The next step is to implement them in a standardized format.

The list of pre-analytical factors should cover the entire sample pathway in the biobank. All users of the biobank will use it, so it must be explicit and clear. In response to this demand, the ISBER developed the Standard Pre-Analytical Code (SPREC), *a way of coding current and future protocols*. The first objective is to annotate the sample so that researchers know the history of protocols that may influence the analytical results. Then, the second objective is to adopt these annotations routinely and include them in the biobank's quality management system. Lehmann et al. have identified the main pre-analytical factors that can impact the quality of a *clinical fluid or solid biospecimen* [LEH 12]. Let us take the example of a liquid fluid and see how the collection, processing and storage data are recorded according to the SPREC.

The first two parameters identify the type of sample and the initial collection tube. For example, whole blood is coded "BLD" (blood), plasma "PL1" after a single centrifugation or "PL2" after a second centrifugation and serum "SER". The container that receives the sample specifies the presence of additives or anticoagulants, for example, the tube containing EDTA is coded "SED" for sodium EDTA.

After identification of the sample and its collection tube, the centrifugation protocol is described by centrifugal force, time, temperature and presence (or absence) of a brake. The code contains a single letter. For example, "A" means a centrifugal force of less than 3,000 g, for 10–15 minutes, at room temperature and without a brake. If centrifugation is not immediate, the time before – and possibly after – centrifugation is also specified. For example, in post-centrifugation, "A" corresponds to a time of less than 1 hour at a temperature between 2°C and 10°C.

Finally, the storage code specifies the type of container used, its volume and the storage temperature. For example, "A" corresponds to a 0.5–2 mL polypropylene tube stored between –85°C and –60°C.

Launched in 2009 and regularly updated by the ISBER, the SPREC has been enhanced with new methods. Version 3.0, released in 2018, details the tree of choices for the seven elements comprising the SPREC code, as listed in the work of Betsou et al. [BET 18]. According to a recent survey on quality in European biobanks undertaken by the BBMRI, the SPREC was implemented in half of biobanks and was planned to be in the remaining biobanks surveyed [LIN 19]. However, this tool will be less and less often used with the implementation of the new ISO 20387 standard, which allows biobanks to opt for accreditation or certification of their activities as a way to formalize their skills. This survey also provides another piece of information about biobanks' quality intentions: if they were previously preparing to implement both ISO 9001 and ISO 20387, the majority are now aiming for ISO 20387. This biobank-specific standard aims at standardizing the entire process followed by the biospecimen, from collection to analysis, during the crucial steps of the pre-analytical phase.

1.3.2. *Protection of the human person and personal information*

Technical harmonization paves the way for standardized operating procedures in biobanks. This improves the robustness of results and accelerates the development of medical research. Beyond the scientific interest, it should be kept in mind that biobanks are *collections of human material* (fluid, tissues, cells or DNA). The whole activity of the biobank is based on this "human" component, which is one of the biobank's priorities, and it must guarantee biospecimen donors their individual rights and protect them. In contrast to technical standardization, the ethical and legal framework of biobanks is far from being standardized with regard to the different actors and their varied interests.

Some Northern European countries have adopted a specific legal framework for biobanks. The Icelandic Act on Biobanks no. 110/2000 was the first European text to regulate the practices of Icelandic biobanks in 2000. Then, in Estonia, ethical, legal and societal issues were addressed in a special law: the Human Genes Research Act of Estonia. Sweden followed with the Swedish Act Biobanks in Medical Care (SFS 2002:297) and Finland biobank legislation (Biobank Act 688/2012) came into force in 2013. In the absence not only of a local law specific to the activity of biobanks, but also European regulation, biobanks must refer their activities to the general law. These activities raise a number of ethical questions about the patient and the biospecimen. How can the patient's wishes be respected? How can their agreement be obtained? Who owns the samples? How can they be distributed? We try to answer these questions by outlining the legal framework of biobanks in France. As there is no specific law, a complex system composed of a set of texts, laws and decrees of the French Public Health Code (*Code de la Santé Publique* (CSP)) takes up the international principles of ethics and research.

1.3.2.1. *The complex legal framework of French biobanks*

Promulgated in 1988, the *Huriet–Sérusclat law* is the founding text for the protection of persons who take part in biomedical research. Modified and then simplified in 2004, law no. 2004-806 relating to public health remains an essential turning point in the history of the development of clinical trials in France and in Europe at the end of the 20th century [FRE 04a]. In order to respond to the protection of participants and the requirement for transparency of clinical data, the law sets out the essential points consisting of the clinical trial sponsor, the personal protection committee and the patient's consent:

– the "*sponsor*" concept identifies the individual or legal entity that takes the initiative for medical research. They manage the implementation, the management and the financing of the study. The sponsor entrusts the study to an experienced "investigator", a natural person who supervises the project. For multicenter trials, several investigators are recruited;

– clinical research protocols are then submitted to an advisory committee for the protection of individuals. The CPP, an independent body with a legal personality, is composed of representatives of patient associations or users of the health system (approved under article L. 1114-1 of the CSP). Its members are appointed by the legal representative of the state in the region where the CPP is located, which may also have several CPPs. The favorable opinion of the CPP is mandatory to start a research project;

– the individual included in the research project gives *free, written and informed consent*. Persons participating in the research may be healthy volunteers or sick persons. They must be affiliated with a social security system and must undergo a medical examination prior to the research, the results of which will be transmitted to them;

– non-respect of these rules can result in sanctioning by the Penal Code (articles 223-8 and 223-9).

In Europe, the ethical and regulatory framework of clinical research is governed by *Directive 2001/20/EC*, which harmonizes the rules on the safety and vigilance of therapeutic trials between member states. This European directive strengthened the rights of participants and created the database of serious adverse events (Eudravigilance). Directive no. 2001/20/EC was transposed into the French Public Health Code in *law no. 2004-806 on public health policy*. From then on, the authorization of clinical trials has been given by the ANSM after the mandatory favorable opinion of the CPP. Internationally recognized good clinical practices (E6: good clinical practice consolidated guideline) are published by the International Council for Harmonisation of Technical Requirements for Pharmaceuticals for Human Use (ICH). Ensuring quality, reliability and harmonization of all documents essential to clinical research are the main objectives of this text, which defines the chain of responsibilities from the sponsor to the investigator and the coordinator designated by the clinical research investigator.

In 2004, the Huriet–Sérustat law was completed by the revisions of the law relating to bioethics and the so-called "CNIL 2004" law [FRE 04b]. *Law no. 2004-800 of August 6, 2004, relating to bioethics* establishes the legal rules organizing the obtaining of biological samples of human origin or health information for medical or scientific purposes. It reinforces the rights of the individual regarding genetic studies and the use of elements and products of the human body. The State created the *Agence de biomédecine*, whose mission is to control research in the fields of transplantation, reproduction, embryology and human genetics. The same year, the law on the protection of individuals with regard to the processing of personal data (law no. 2004-801 of August 6, 2004) revised the framework for research on "human beings". Since 2004, the CNIL law and the bioethics law have been revised.

The law no. 2012-300 of March 12, 2012, relating to research involving the human person, known as the *Jardé law*, was implemented in 2016 by decree no. 2016-1537 of November 2016 [FRE 12]. This decree specifies the modalities for conducting research involving the human person. It defines two types of research – interventional and non-interventional – and replaces the notion of "biomedical research" with "research involving the human person". It determines the missions of the French National Commission for Research (*Commission nationale des recherches*) involving the human person, including the coordination of CPPs in conjunction with sponsors, the ANSM and the CNIL. In 2017, a new decree was published in the *Journal officiel* (decree no. 2017-1549 of November 8, 2017) relating to the conservation and preparation for scientific purposes of human body elements. It applies to the activities of conservation and preparation for scientific purposes of tissues and cells of the human body:

– for the specific needs of an organization's research programs, which requires a declaration to the Minister of Research (see article L. 1243-3 of the French Public Health Code);

– with a view to their transfer for scientific use, which requires authorization issued by the Minister of Research (see article L. 1243-4 of the French Public Health Code).

These declaration procedures applicable to the exercise of activities of conservation and preparation of elements of the human body for scientific purposes consequently modified the French Public Health Code. The modifications made by the decree were updated by the application for the management of the conservation of human body parts (Codecoh) of the Ministry of Higher Education, Research and Innovation. The declaration and authorization request forms are available online. In both cases, the competent territorial CPP must also be notified. For a health establishment, the declaration – or the request for authorization – is also sent to the director of the regional health agency (*Agence Régionale de Santé* (ARS)) with territorial jurisdiction. For an army hospital, the declaration – or the request for authorization – is also addressed to the Minister of Defense.

Research activities on embryos and embryonic stem cells are subject to a specific legal framework. The law on bioethics (law no. 2004-800 of August 6, 2004), reformed with law no. 2011-814 of July 7, 2011, was modified by law no. 2013-715 of August 6, 2013, "authorizing under certain conditions research on embryos and embryonic stem cells". The 2019 bioethics bill aims to facilitate strictly supervised research on embryonic stem cells. It proposes the distinction of the legal regime for embryos and embryonic stem cells (article 14 voted on October 4, 2019).

Currently, the conservation of embryos and embryonic stem cells for research purposes requires authorization issued by the biomedicine agency (CSP, art. R. 2151-20). Any organization conserving embryos or embryonic stem cells for research purposes must be able to justify the collection of consent given by the embryos' parents. When embryonic stem cells have been imported, the organization must be able to justify that they have been obtained in compliance with the prescriptions of the Civil Code, with the prior consent of the couple who generated the embryo conceived in the context of medically assisted procreation, and that this embryo is no longer the subject of a parental project, without any payment, in whatever form, having been allocated to the couple (CSP, art. R. 2151-18).

1.3.2.2. *Protection of the biospecimen donor*

The biobank is at the center of a network of rights and of obligations towards the biological sample donor. Upstream, it must manage the rights of individuals by obtaining consent forms, and downstream, it must manage the conditions of access of donors and researchers to the data associated with biological samples. A body of laws and decrees indicates the necessity of informed consent in research involving the human person:

– the law on information technology and freedom, no. 78-17, modified by law no. 2004-801 of August 6, 2004, relating to the processing of personal data;

– law no. 2002-303 of March 4, 2002, relating to the rights of patients;

– law no. 2004-800 of August 6, 2004, relating to bioethics;

– the Jardé law on research involving the human person, decree no. 2016-1537 of November 16, 2016.

Free and informed consent is mandatory for the use of a human sample. According to article L. 1122-1-1 of the French Public Health Code, the rights of individuals must be respected: "No research may be carried out on a person without his or her free and informed consent, obtained in writing, after he or she has been given the information provided for in article L. 1122-1."

By signing this document, the individual expresses their willingness to participate in the research project, which will have been explained in a notice attached to the consent form. With this information, the investigator, or the authorized person, explains why the sample provided is essential to the conduct of the project. The investigator answers the donor's questions and both sign the consent form. The donor may withdraw from the study at any time. This consent is mandatory for the use of biological samples and personal data or for genetic research.

Consent to participate in the research project is based on:

– the participant's written consent (inclusion), which is required;

– the information about the research project provided by the investigator, such as the objectives, methods, benefits and risks associated with the study. It is also specified that the participant is free to accept or refuse to participate in the study in question;

– the investigator's prior declaration to the CPP, which is authorized to obtain the participant's consent.

Before the study begins, the consent procedure is explained orally to the patient by the investigator. The consent form mentions the patient's rights; it is dated and signed and kept by the investigator. The patient receives a copy which they also keep, as well as the information leaflet.

Authorization is granted in the context of a particular study. However, the participant's sample or associated data may have a different purpose. If the participant does not object to the reuse of their sample, it will be used in a future project (article L. 1211-2, paragraph 2 of the French Public Health

Code). In France, the secondary use of samples is based on the principle of non-opposition. More generally, if the reuse concerns several samples, the scientific manager may request a *global re-qualification of a collection* to the CPP, explaining the objectives of the new research and especially the need to include the samples of unreachable participants. The CPP can then accept the requalification, except in the case of genetic studies, for which individual consent is mandatory, and for the use of germ cells.

When a biobank *transfers a collection* for a new research project, the scientific director must make a declaration to the Ministry of Research, which takes into account the opinion of the CPP. The biobank may transfer its samples by means of a material transfer agreement (MTA) in accordance with articles L. 1243-3, L. 1243-4 and R. 1243-49 of the CSP. This document contains the references and the purpose of the applicant (the identity of the applicant, the structure of the institution, the context of the request, the use of the biological materials, the charges and indemnities, respect of confidentiality).

The principles of the non-opposition of the patient and of requalification, or even transfer, of a collection of biological samples show the limits of the traceability of consent. The future of consent is projected in the form of an electronic version of informed consent. Already regulated by FDA guidelines in the United States but still non-existent in France, *electronic informed consent (e-consent or eIC)* refers to "the use of electronic systems and processes that may employ multiple electronic media, including text, graphics, audio, video, podcasts, passive and interactive Web sites, biological recognition devices, and card readers, to convey information related to the study and to obtain and document informed consent. The information presented to the subject, processes used for obtaining informed consent, and documentation of the electronic informed consent (eIC) must meet the requirements of these and other applicable regulations" [FDA 16]. The 2016 FDA guidance for scientific leaders emphasizes the dynamics of patient interaction and information enrichment. It offers traceability of the sample in real time: the clinician controls the receipt of documents and evaluates the correct understanding of the information by the patient. For their part, manufacturers monitor the use of samples and data.

As well as requests for authorization and declarations to the competent authorities, the exchange of information between research teams is mainly carried out electronically. These new computer technologies are being integrated into the lives of patients with the implementation of *electronic health records (EHRs)* in most US hospitals and shared medical records in France (*dossier médical partagé*). In clinical research, data collection is recorded in an *electronic case report form (eCRF)*. This contains all the individual information collected during the study: patient identification, study inclusion criteria, informed consent and all medical examination results. It centralizes the individual management of the clinical study, which can potentially be accumulated for all patients in order to conduct a clinical trial that is largely numerical and respects the ethical and regulatory bases. Obviously, biological analysis samples are obtained in laboratories close to the patient's home and medical examinations are carried out by the attending physician. Over the past 20 years, digital transformation has made its entrance into the hospital and more broadly into healthcare with the rise of e-health.

1.3.2.3. *Protection of the data associated with the sample*

French law provides a protective framework for personal data with the law on data processing, files and freedoms (law of January 6, 1978, as amended), which the CNIL is responsible for enforcing. Article 8 of the law stipulates:

> It is prohibited to collect or process personal data that reveal, directly or indirectly, the racial or ethnic origins, political, philosophical or religious opinions or trade union membership of persons, or that relate to the health or sex life of such persons.

Computerized systems centralizing a vast amount of clinical, family, environmental and personal information must be able to ensure the *confidentiality* of all data associated with the donation of biological samples. The constitution of a collection of human biological samples for genetic or non-genetic research purposes is governed by articles L. 1131-4, L. 1243-3 and L. 1243-4 of the French Public Health Code. Once the collection project has been declared to the competent administrative authority, the latter ensures that the conditions of constitution, conservation and exploitation of the collection present sufficient guarantees to ensure the proper use, security and confidentiality of the data collected.

Upon receipt, the biospecimen starts its journey in the biobank with a coded identification. The management of the samples requires *the anonymization of data*. Absolute anonymization is not consistent with the principle of returning clinical information to the patient. Pseudonymization is reversible and enables re-identification of the patient. The person's identity information (surname, first names and social security number) is coded by a key or encryption code assigned to all samples from the same individual. This coding facilitates the longitudinal follow-up of the patient, which can extend over several years during an epidemiological study.

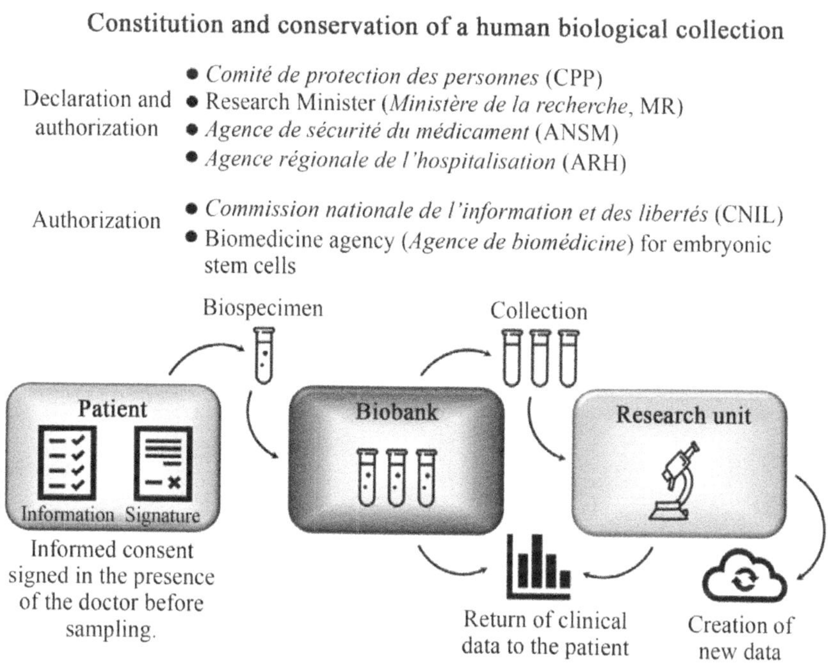

Figure 1.8. *The legal framework for biobanks for the conservation of a collection of human samples*

COMMENTARY ON FIGURE 1.8.– *After the favorable opinion of the CPP, the person in charge of the biobank declares the conservation of the collection to the Ministry of Higher Education, Research and Innovation and to the ANSM, as well as to the French Regional Agency for Hospitalization (Agence Régionale de l'Hospitalisation (ARH)). The creation of new data from research is submitted for authorization to the CNIL. In the case of embryonic stem cells, authorization to store the collection is requested from the Biomedicine Agency (Agence de Biomédecine).*

At the European level, the new *General Data Protection Regulation* (GDPR) reinforces the principles of data protection and the rights of citizens, repealing Directive 95/46/EC (Data Protection Directive) [EUR 16]. This regulation applies to activities of the biobank both upstream, the missions related to the collection of personal data, and downstream, the creation of new data from and associated with the samples [CHA 16]. The processing of personal data must respect in particular law no. 78-17 of January 6, 1978. relating to "information technology, files and freedoms", consolidated version from July 23, 2019. The CNIL must be consulted before any data processing to ensure that security and confidentiality are respected, as well as the proper use of personal data, especially in the case of genetic data.

The third pillar of medical research, after the management of biological samples and complex data, is the regulatory and legal framework that ensures respect for the rights of individuals. Guaranteeing the protection of patients is a reinforced priority for biobanks, which are subject to the accelerated pace of scientific and technological innovations.

2

Biobanks in the Digital Age and Precision Medicine

Artificial intelligence is shaking up the world of medical imaging.

[AYA 18]

New technologies are at the heart of biobank growth and health data management. Next-generation biobanks will have to meet the challenge of digital data. With high throughput technologies such as omics combined with biological, molecular or imaging approaches, the integration of massive data should accelerate medical research. This megadata, in the form of numerical values, images and behavioral and family data, hosted in dematerialized databases, is grouped under the term “Big Data”. In practice, the analysis of data from a large number of patients is evolving into the analysis of a large volume of data from a single patient, referred to as “Fat Data”. The pyramidal increase in the volume of data with the number of patients requires computerized means to process them. Thanks to artificial intelligence and the computing power of computers, it is now possible to create algorithms to obtain automatic and precise answers.

Biobanking in the digital age is distinguished by:

– integration of medical imaging data;

– quantitative analysis of the imaging essential to the diagnosis and decision support of the physician;

– the standardized and computerized management of international expertise biobanks.

2.1. Medical imaging biobanks

Medical imaging is a major resource in epidemiological research. Traditionally, longitudinal cohort studies have been based on the clinical monitoring of risk factors for the development of various diseases. With large-scale prospective cohorts, new risk factors have been identified. They are still monitored today at the individual level, but monitoring them at the general population level is much more complex.

In the last decade, imaging techniques have become more widespread in population-based cohorts with the aim of detecting early signs of disease. New imaging biomarkers have been associated with the developmental stage of cardiovascular disease, for example, in the Framingham heart study described in the previous chapter and the Rotterdam scan study detailed in the next section [SPL 07].

Medical imaging biomarkers have two major advantages in the follow-up of large prospective cohorts:

– detecting early signs before clinical symptoms allows preventive measures and treatments to be implemented at the very beginning of the disease rather than at the advanced stage;

– the quantitative approach, which involves measuring a volume or a diameter, brings a statistical power that is superior to clinical records.

Today, medical imaging with magnetic resonance imaging (MRI) techniques make it possible to visualize an organ and analyze its morphology and function without using contrast agents. It is also possible to explore the entire body during a routine examination. These developments have been applied in various explorations, such as into the distribution of fatty tissue, brain and heart functions and those of abdominal and thoracic organs.

Population-based cohorts, with imaging databases of thousands and tens of thousands of participants, are included in several international research projects. Three studies are detailed below: the UK Biobank imaging study, the Rotterdam scan study and the German National Cohort.

2.1.1. *The UK Biobank prospective medical imaging study*

Obesity is spreading rapidly in Western societies and is causing adverse health effects. For the past 30 years, studies on obesity have explored its consequences, such as cardiovascular diseases and certain types of cancer, in order to better fight them. The regulation of weight has been explained by complex physiological, neuronal and molecular mechanisms.

More recently, a new question has been formulated in reverse. What is the effect of obesity on cognitive phenomena and brain health? Excess weight and obesity have been shown to accelerate cognitive decline and dementia, highlighting a chronic inflammation common to both clinical pictures [DAN 19]. The effects of excess weight and obesity on the brain are poorly documented. They require robust studies including hundreds or thousands of participants representative of the population. Only clinical follow-up brain imaging with in vivo MRI allows for a large-scale study protocol.

The UK Biobank has launched a large medical imaging study. Its goal is to build the world's largest collection of MRI scans of vital organs with over 100,000 participants [UKB 20]. This will allow researchers to see how organ structure and function relate to disease development. Nearly 50,000 volunteers have agreed to participate and help further study vital organs such as the brain, heart, tissues and bones using various techniques:

– brain MRI to understand the realization of certain tasks and the connection between the different parts of the brain;

– cardiac MRI to measure the size of the heart chambers and blood vessels and their changes in size during heartbeat;

– whole body MRI to estimate the distribution of body fat in different areas of the body;

– ultrasound examination of the carotid artery and neck vessels.

The study on obesity and brain structure appeared in the journal *Radiology* in 2019. Dekkers et al. examined MRI brain images from the UK Biobank, one of the few to have integrated radiological examinations along with biospecimen collection during longitudinal follow-up [DEK 19]. Dekkers and colleagues at Leiden University Medical Center in the

Netherlands analyzed brain images of participants and showed that obesity can affect brain structure.

Begun in 2006, the British Biobank Prospective Population Cohort Study recruited 503,325 volunteers aged 40–69 years for five years to identify genetic and environmental risk factors associated with disease development. Of the participants, 12,087 volunteers, 52% of whom were women, gave their consent for medical MRI scans, which were performed using 3 Tesla multiparametric MRI, the gold standard imaging technique for detecting anatomical details of brain structures. The cross-sectional images show the white and gray matter of the participants. Simultaneously, excess weight and obesity were assessed for each participant using routine measurements such as body mass index (BMI), body and fat mass estimated by impedancemetry, waist circumference and hip circumference.

Thanks to MRI, the sizes of the different structures of the brain were estimated. The total volume of the brain was measured as well as the volumes of the different brain regions: gray matter and white matter. The gray matter located in two places, on the surface of the cortex and deep inside, was measured according to its cortical volume and its subcortical (internal) volume.

The BMI and body fat percentage results together give an idea of the distribution of profiles in the overall population. The mean BMI value of 26.6 kg/m^2 ± 4.4 (standard deviation) indicates a significant variation. This is the sign of a heterogeneous population, a sample composed of thin, normal, overweight or obese people. The percentages of fat mass confirm this variation in men (24.4 % ± 5.5) as well as women (35.5% ± 6.5). The different profiles represent a fraction of the overall population.

In men, the percentage of fat mass is negatively associated with the total volume of subcortical gray matter (thalamus, hypothalamus, etc.). In women, the percentage of fat mass is negatively associated only with the volume of subcortical gray matter of the globus pallidus. These results show that obesity is associated with a smaller volume of deep gray matter and a change in the geometric organization of the white microstructure. Dekkers et al. also observed that these associations are different between men and women, showing that gender affects the correlation between the percentage of fat mass and the size of brain structures.

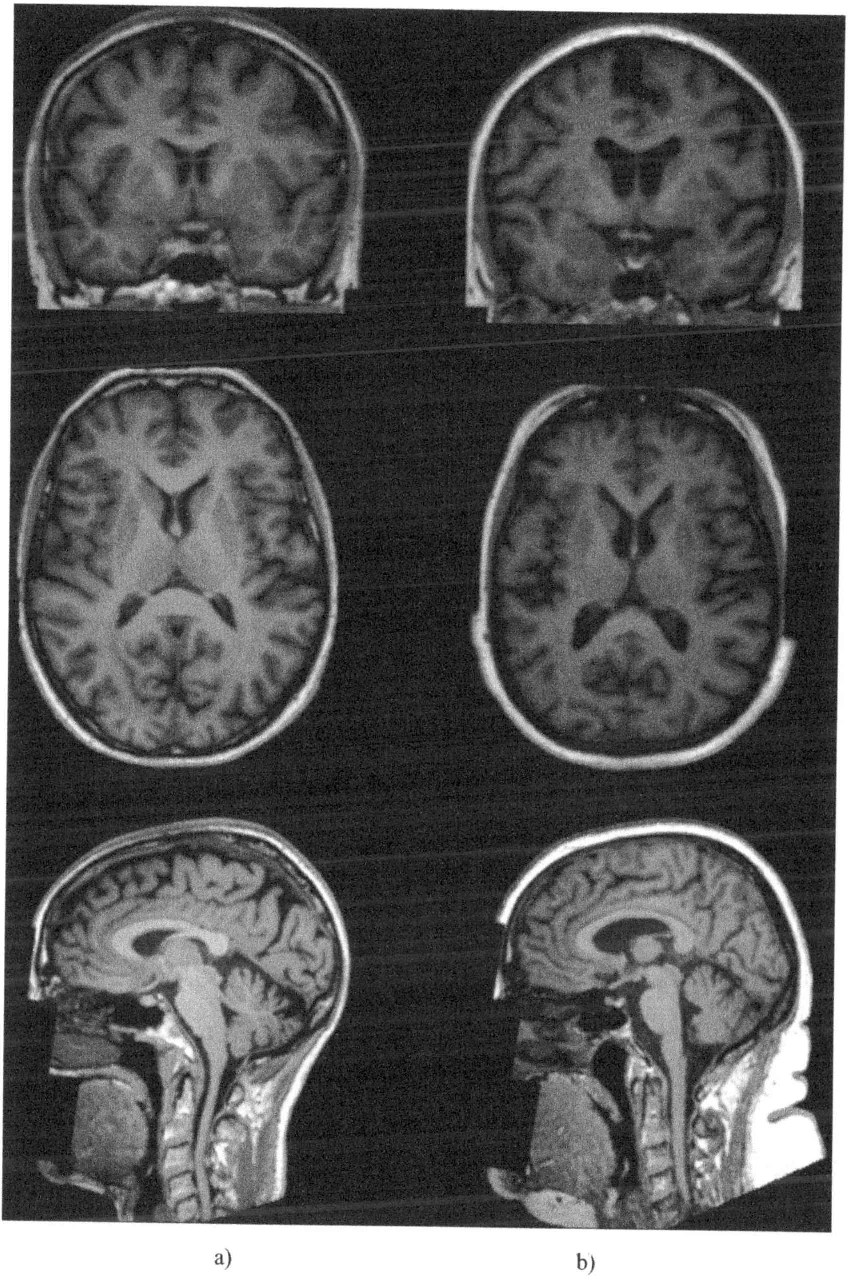

Figure 2.1. *Two examples of magnetic resonance imaging (MRI) of the brain in coronal (top), axial (middle) and sagittal (bottom) planes*

COMMENTARY ON FIGURE 2.1.– *The two participants in the UK Biobank study were two women aged 65 years. The percentage of fat mass reached (a) 13% for the first and (b) 49% for the second. MRI showed a smaller volume of subcortical gray matter for the individual with the higher body fat percentage [DEK 19]. Photo credit: Radiological Society of North America.*

According to these results, a decrease in gray matter volume suggests a loss of neurons and a change in white matter affects the transmission of signals in neural networks. Knowing that the subcortical gray matter plays a role in the reward system, these changes could explain the difficulty that obese people have in controlling their weight. By considering total body fat, this study did not distinguish the role of different adipose tissue deposits, including visceral tissue, which is a risk factor for cardiovascular disease and diabetes.

"MRI has been shown to be an irreplaceable tool for understanding the link between neuroanatomical differences of the brain and behaviour", concludes Ilona Dekkers, the first author of this study.

2.1.2. *The Rotterdam prospective imaging study*

The Rotterdam study is a prospective cohort study started in 1990 in the city of Rotterdam in the Netherlands. Its primary objective is to study the aging of the population by targeting cardiovascular, endocrine, liver, neurological and pulmonary diseases. In 2008, the Rotterdam cohort had 14,926 participants aged 45 years and older. Since 2016, additional volunteers over the age of 40 have joined. In 2018, major results were published by the Ikram et al. adjusting the protocol to the new objectives [IKR 17]. In total, more than 1,700 publications have emerged from this cohort [IKR 20].

Since 1990, participants in the study have filled out a health questionnaire (medication use, medical history) and described their lifestyle habits (smoking, alcohol consumption and physical activity). Every three or four years, they undergo a complete check-up including medical imaging and blood sampling for molecular and genetic analysis. They are monitored for health conditions such as cardiovascular disease (heart failure and stroke), neurodegenerative disease (Parkinson's disease, Alzheimer's disease, dementia, depression and other anxiety disorders), eye disease (macular degeneration and glaucoma), diabetes and osteoporosis.

The Rotterdam study implemented medical imaging in the follow-up of the general population (population imaging unit). Disease imaging biomarkers are detected at a presymptomatic stage by two non-invasive techniques, MRI and multidetector-computed tomography (MDCT). They characterized atherosclerosis to better understand the atherosclerotic process and identified risk factors for cardiovascular disease by measuring coronary calcification and characterizing carotid plaque with MRI.

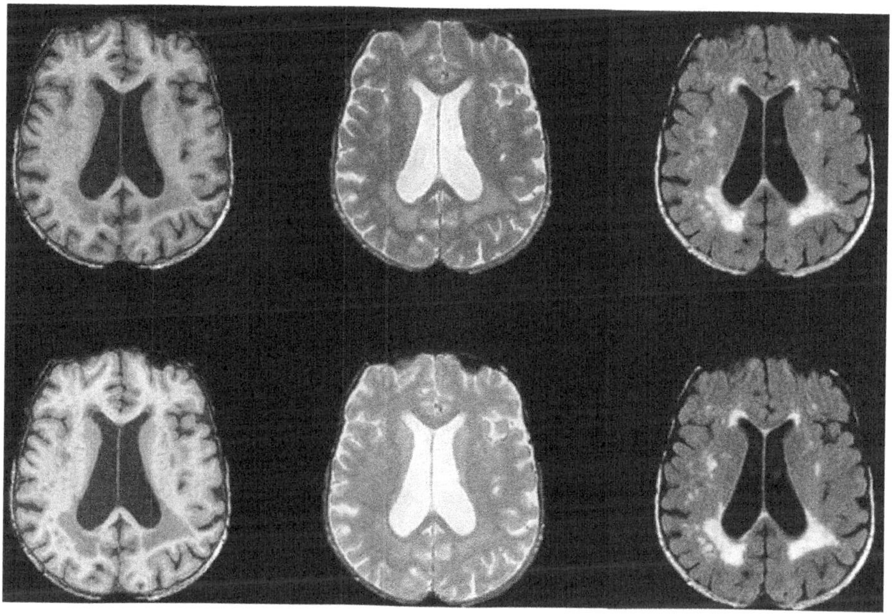

a) b) c)

Figure 2.2. *An example of a brain MRI with different image segmentations. For a color version of this figure, see www.iste.co.uk/arrighi/biobanks.zip*

COMMENTARY ON FIGURE 2.2.– *(a) T1-weighted MRI, different regions localize gray matter (orange), white matter (green), cerebrospinal fluid (purple) and white matter lesions (red). (b) Proton density-weighted image with identical segmentation, regions are colored the same. (c) Fluid-attenuated inversion recovery (FLAIR)-weighted sequence with white matter segmentation (pink). Data taken from Ikram et al.'s work summarizing key advances and protocol changes in the Rotterdam imaging study from 2016 [IKR 15].*

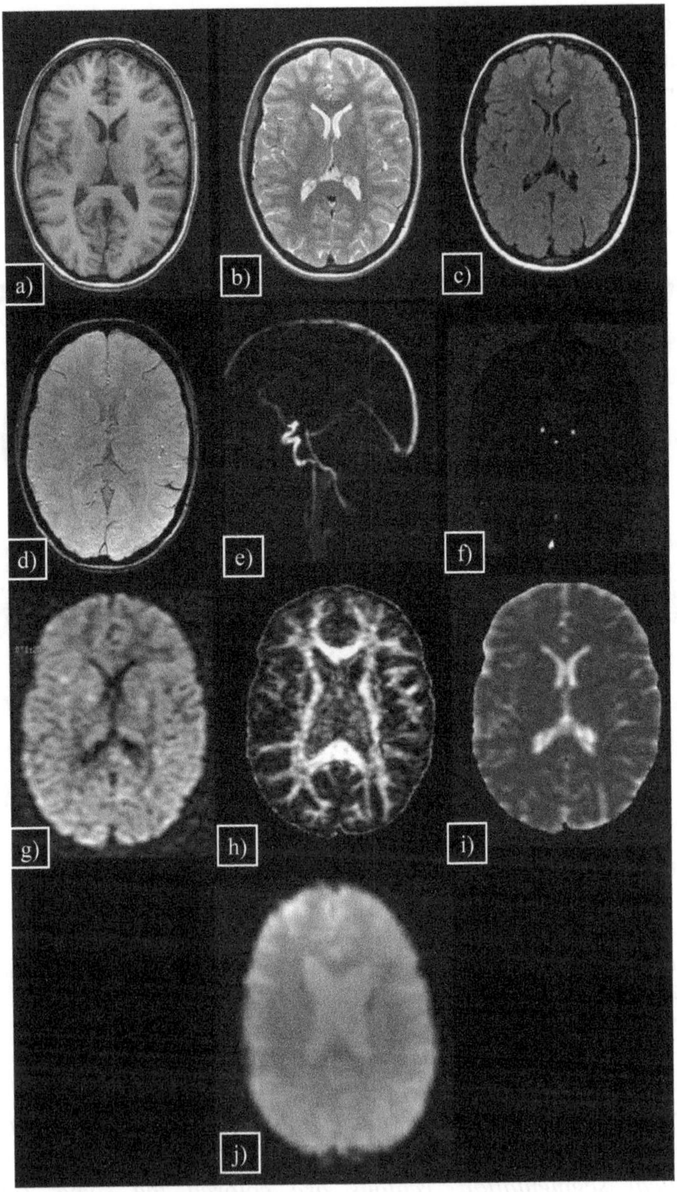

Figure 2.3. *Brain MRI protocols from the Rotterdam imaging study [IKR 15]*

COMMENTARY ON FIGURE 2.3.– *First row: (a) T1-weighted image, more intense white matter appears bright and less intense gray matter dark (cerebrospinal fluid is black); (b) proton density-weighted image (white matter is more proton dense, myelination zone); (c) FLAIR. Second row: (d) T2-weighted image, gray matter appears bright and white matter less intense and dark; (e) sagittal phase contrast slice to quantify blood flow; (f) final blood flow image. Third row: (g) diffusion tensor image; (h) fractional anisotropy map; (i) mean diffusivity map; (j) resting state functional MRI.*

At the start of the pilot study in 1991, a sample of 111 participants examined by brain MRI showed white matter abnormalities [BRE 94]. In 1995, 563 participants underwent this examination. Since 2005, brain MRI has been routinely performed for participants in the Rotterdam study who do not have any contraindications. There is a 1.5 Tesla MRI entirely dedicated to the Rotterdam scan study.

General population imaging work has identified a predictive model for stroke risk. Evans et al. followed 4,259 volunteers with no history of stroke and performed brain MRI from 2006 to 2011 [VAS 16]. The average age was 63.6 years and 55.6% were women. Follow-up was extended to 2013. During the study period, 58 cases of stroke were counted. White matter impairment was found to be associated with stroke risk. Evans et al. concluded that stroke is predicted by the prevalence of vascular lesions and also by fine alterations in white matter microstructure. This new parameter was added to the reference model, the Framingham score. The latter was used to identify the main risk factors for cardiovascular disease, such as high blood pressure, high blood cholesterol, smoking and obesity, in the Framingham epidemiological study described in the previous chapter.

This new predictive model improved on the Framingham model by adding a new factor to the existing parameters related to the alteration of white matter integrity.

2.1.3. *The German National Cohort*

The German National Cohort is a prospective cohort studying the development of chronic diseases in the general population since 2014 [NAK 20]. The major objective is to identify risk factors associated with cardiovascular, neurological, respiratory, infectious and muscular diseases, diabetes and cancer. With such a broad objective, whole-body MRI provides a technique for exploring all organs in a comprehensive and thorough examination for each participant. As the cohort represents the entire German population, volunteers scattered throughout the country were accommodated at several examination centers. Procedures were standardized in this multicenter study, including operating protocol, quality assurance and data management.

The imaging repository plays a central role in the development of this study in three ways:

– determining predictive values for the incidence of chronic diseases;

– determining the prevalence of preclinical disease in the general population;

– developing screening methods for asymptomatic individuals.

The German National Cohort Study is an interdisciplinary effort involving several research institutes. Eighteen study centers were distributed throughout the country in order to gather a representative sample of 200,000 participants, including men and women aged 20–69 years. Every four to five years, the volunteers undergo a complete medical examination, including cognitive and cardiac tests as well as biological sampling of blood (plasma, serum, DNA, RNA and red blood cells), saliva, nose and feces. A more thorough examination is planned for 20% of the participants, including an electrocardiogram, a glucose tolerance test and ophthalmological, olfactory and muscle tests.

A subgroup of 30,000 participants from the 200,000 volunteers in the cohort were recruited to undergo whole-body MRI scans at one of five dedicated MRI study centers in the states of Bavaria, Berlin, North Rhine-Westphalia, Mecklenburg-Western Pomerania and Baden-Wurttemberg.

The sample size was calculated to detect early signs of disease in the general population. The sample size is large enough to detect common diseases such as fatty liver (20% prevalence) as well as less common diseases such as cerebral microhemorrhage (5% prevalence) and chronic pancreatitis (2.5% prevalence). After 10 years of follow-up, 2,000 cases of cancer, 5,000 cases of cardiovascular disease and 2,500 cases of death from other causes were detected.

The examinations are performed on 3 Tesla multiparametric MRI equipment that is identical in all imaging centers. The acquisition time is approximately one hour. Morphological and functional imaging is undertaken globally for several organs and parts of the body, including the brain, heart, thorax and muscles, with a standard protocol:

– brain ımaging allows for analysis using three approaches:

- a T1-weighted image to measure the volume of brain structures,

- a T2-weighted image and FLAIR to detect white matter abnormalities, mainly resulting from cardiovascular disease,

- a resting state functional MRI to explore the functional organization of the various areas of the brain;

– cardiovascular imaging assesses left ventricular function and mass, and a T1-weighted image detects changes in cardiac muscle tissue;

– thoraco-abdominal imaging measures the volume of the organs with a weighted image:

- in T1 with the Dixon technique, which differentiates adipose tissue from lean tissue. The quantification of adipose tissue is performed in different deposits, such as visceral and epicardial adipose tissue, which represents a risk factor for metabolic and cardiovascular diseases. The multi-echo sequence determines the accumulation of fat in the liver, which can be a prognostic marker of liver disease,

- in T2, which analyses the free water content present during inflammation or edema;

– musculoskeletal imaging sometimes overlaps with thoraco-abdominal imaging. However, it mainly targets osteoarthritis and inflammatory joint diseases. *The protocol uses 3D isotropic* fast spin echo *sequences* with several flip angles, which allows for the exploration of lesions at the level of articular cartilages.

The special report in *Radiology* presents the progress of the whole-body imaging project. Bamsberg et al. explain the background, purpose and technical protocols of this prospective population-based study [BAM 15]. Implementing a large MRI study in the German national cohort requires coordination of the five imaging centers over four years to reach the expected 30,000 participants. In conclusion, Bamsberg summarizes that "Population-based whole-body MR imaging as part of the German National Cohort will serve to compile a comprehensive image repository, will provide insight into physiologic variants and subclinical disease burden, and has the potential to enable identification of novel imaging biomarkers of risk" [BAM 15].

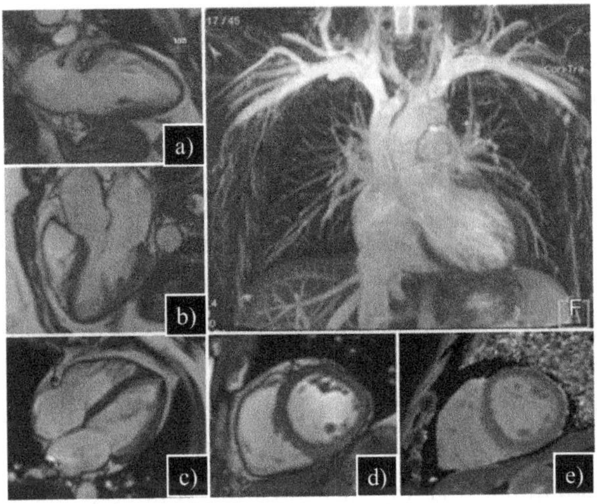

Figure 2.4. *Cardiovascular MRI*

COMMENTARY ON FIGURE 2.4.– *(a)–(c) MRI for longitudinal section; (d) transverse; (e) T1-weighted image from FLAIR and MRI thoracic angiography. Images from the* Radiology *special report presenting the whole-body imaging project of the Bamsberg team [BAM 15].*

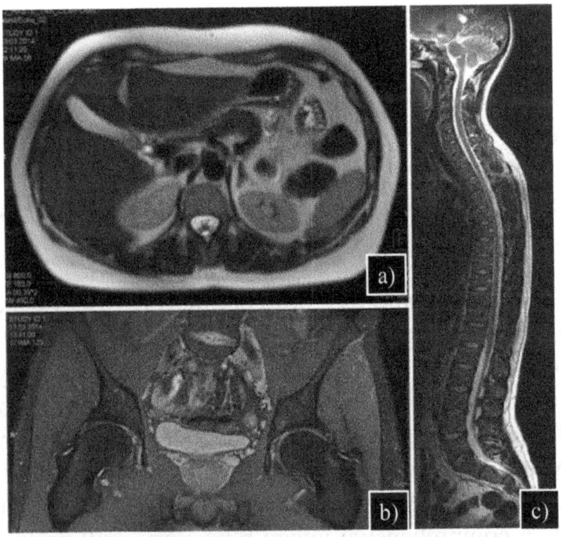

Figure 2.5. *(a) T1-weighted MRI of the thorax and abdomen and (b) musculoskeletal MRI of the pelvis and (c) the spine*

COMMENTARY ON FIGURE 2.5.– *Images from the* Radiology *special report featuring the Bamsberg team's whole-body imaging project [BAM 15].*

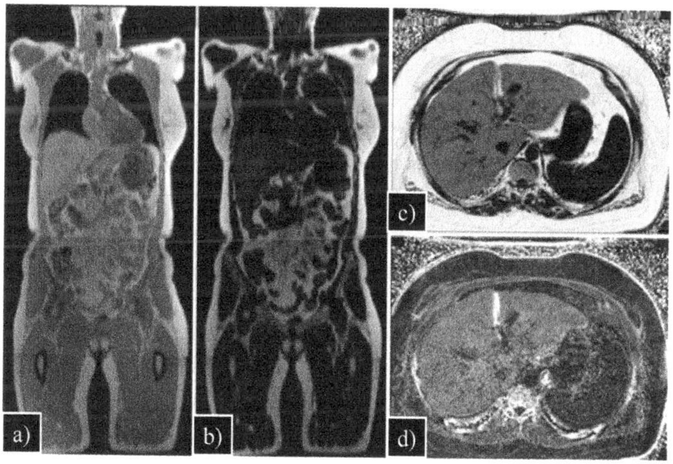

Figure 2.6. *Thoraco-abdominal MRIs. (a) T1-weighted image from neck to the knees; b) fat mass according to the Dixon 2-point technique; c) multi-echo image with six fat mass shade echoes; d) the liver image*

COMMENTARY ON FIGURE 2.6.– *Images from the* Radiology *special report featuring the Bamsberg team's whole-body imaging project [BAM 15].*

2.1.4. *The European medical imaging biobank project*

Until recently, MRI data were not included in biobanks. Rarely, projects such as the UK Biobank, the Rotterdam study and the German National Cohort have included MRI scans in the epidemiological follow-up of population cohorts.

In 2014, the European Society of Radiology (ESR) launched a European plan for medical imaging [EUR 15]. It defines the imaging biobank as "organised databases of medical images, and associated imaging biomarkers (radiology and beyond), shared among multiple researchers, linked to other biorepositories" [EUR 20a]. This action plan calls for increased harmonization of safety, training, research and e-health in the field of radiology. It is in line with the European health and research program Horizon 2020.

Most biobanks massively collect omics data while imaging data are rarely collected. Yet genotype information combined with phenotype plays a key role in personalized medicine. The era of personalized medicine and biobanks has transformed the one-size-fits-all healthcare offer into an individual approach from diagnosis to treatment.

Imaging biomarkers are essential in personalized medicine, not only in detecting disease in patients, but also in tailoring patient-specific treatment. The identification of new biomarkers involves the analysis of large cohorts and massive data. Only biobanks provide access to large-scale samples offering a wide range of genetic, proteomic and transcriptomic analyses.

Imaging biobanks are infrastructures with large digital storage and processing capacities. High-performance computing resources are required to perform image processing and comparison. All information extracted from medical imaging is considered as the phenotypic expression of the patient and can be linked to the genotype.

In the ESR's action plan, two main ideas stand out:

– imaging biomarker data are shared in a network with biomedical banks;

– traditional and imaging biobanks are moving forward together in the context of personalized medicine.

The collaboration signed in 2015 between the ESR and the BBMRI-ERIC marked the launch of a European medical imaging biobank. Through the 2014 European Medical Imaging Action Plan, the ESR highlighted the need for a structured imaging data repository to facilitate personalized medicine, clinical trials and new drug evaluation.

The main objective of this collaboration is to promote the visibility of imaging biobanks and to build a European imaging biobank infrastructure linked to existing biobanks. To achieve this, the ESR advocates:

– supporting efforts to standardize and validate technical and clinical imaging biomarkers;

– improving interoperability, standardization and data management and ensuring a harmonized approach to data quality assurance;

– integrating imaging data with omics databases.

The two organizations have worked together on the connection between the minimum information about biobank data sharing (MIABIS 2.0) and the digital imaging and communications in medicine (DICOM) standards. The information common to both registries is implemented in the BBMRI-ERIC 2.0 and subsequent versions. In addition to the imaging data, the BBMRI-ERIC 3.0 version includes quality indicators.

In November 2019, the European Union funded the Euro-Bioimaging project, which sets the legal, financial and technical framework for European bioimaging research infrastructures [EUR 20b]. In general, the objective is to provide scientists with open access to facilities, expertise, training and data services in bio-imaging. As a European research infrastructure consortium (ERIC), Euro-Bioimaging is formed from a network of medical imaging experts in 15 countries (Austria, Bulgaria, Czech Republic, Denmark, Finland, France, Hungary, Israel, Italy, the Netherlands, Norway, Portugal, Sweden, the United Kingdom and Belgium).

With this green light, Europe extends the scope of biobanks to radiology, adding a third pillar to their activities. In addition to the collection of biological samples, which was its first activity, the biobank combines high quality samples with data from biochemical, genomic and clinical analyses, which is the strength of precisely documented collections. This second pillar has been completed by the storage of medical images. The development of image processing software has automated the reading of images and accelerated the identification of new biomarkers.

2.2. Radiomics data powered by digital technology

The role of medical imaging, initially reserved for diagnosis, has become central in the context of personalized medicine. Imaging information has complemented the results of biological, clinical and genomic analyses.

Computed tomography is the classic anatomical imaging technique, with a resolution of up to 1 mm^3. This routine examination follows cancer patients at every stage of the treatment process. From prevention, patients with risk factors are monitored. The early diagnosis stage identifies the disease and then estimates the severity of it (prognosis). Finally, the course of treatment is evaluated in terms of efficacy and toxicity. CT scans are used to evaluate structures but do not provide information on their functionality or molecular composition. In contrast, functional imaging explores physiological processes using dynamic contrast-enhanced magnetic resonance imaging (DCEI), providing in vivo vascular information, for example.

The study by Lambin et al. defined the concept of radiomics as the large-scale extraction of information from medical images. The quantitative analysis of biomarkers in imaging has established databases that are essential for the detection and treatment of diseases. For a single patient, modern medical imaging is multiparametric, characterized by digital information from several biomarkers which is quantified as: diameter, volume, density measurement in CT, signal intensity in MRI, normalized fixation value in proton emission tomography, contrast enhancement (in MRI, CT or ultrasound), perfusion parameters (blood flow, blood volume, mean transit time and permeability), tissue elasticity in elastography, tissue texture and morphological structure.

Over the last 20 years, medical imaging has developed thanks to advances in several fields [LAM 12]:

– new imaging equipment and new contrast agents: CT scanners that used to analyze a single slice plane now have the possibility of combining several slice planes. Contrast agents injected into the body are used as specific metabolic indicators, such as for hypoxia;

– standardized protocols allow quantitative analysis: classical radiology is a qualitative approach, but standardization of protocols transforms radiology into a qualitative discipline, applicable in multicenter studies;

– new image analysis software: this allows quantification and standardization, which improves the performance of automated detection in the case of lung diseases and mammograms.

Radiomics entails the more or less automated extraction of a large number of quantitative imaging parameters. All this information is considered as the expression of the patient's phenotype and can be correlated with the gene expression profile, hence the term radiogenomics. The radiomics process is based on the collection of data after automated extraction of a very large number of descriptors (more than 200) from a medical image. The hypothesis is that automatic quantitative analysis by software provides better information than visual recognition by the radiologist.

As a first step, images are collected in large numbers during routine clinical examinations. All images are acquired in a high quality and standardized manner. The segmentation step is performed automatically or by an experienced radiologist. It involves defining the region of interest, for example, the macroscopic area of a cancerous tumor. From this region, the imaging parameters will be extracted. These descriptors represent intensity levels, texture features, shape descriptors and size measurements. After this extraction, the image descriptors undergo a selection step according to the technical quality. Finally, the selected descriptors are analyzed in correlation with the parameter of clinical interest. The final goal is to perform a stratification by incorporating imaging biomarkers in predictive models and to evaluate their added value compared to classical indicators.

Radiomics has opened the door to a multitude of application areas. We will see how quantitative imaging has replaced expert observation in radiology. The processing of large numbers of images has required new computer tools for the automatic reading of medical images. These software packages have learned the expert's reasoning and identified new biomarkers.

2.2.1. *A multitude of application areas for biomarkers*

Biomarkers are critical to the development of drugs and medical devices. The basic definitions and concepts behind them are collected in the Biomarkers, Endpoints and Other Tools (BEST) document produced by the US FDA and the NIH [FDA 16].

According to the BEST Glossary, a biomarker is defined as:

> A defined characteristic that is measured as an indicator of normal biological processes, pathogenic processes, or biological responses to an exposure or intervention, including therapeutic interventions. Molecular, histologic, radiographic, or physiologic characteristics are types of biomarkers.

This broad definition can be used in a multitude of fields of application [CAL 18]. Some biomarkers are used during the research and development phases of therapeutic molecules, others are used in routine clinical practice, and finally some condition the prescription of a specific treatment. A wide range of biomarkers are used today, alone or in combination. They are classified in several ways.

The diagnostic biomarker is used to detect lesions and confirm the presence of a disease. This type of biomarker is not only used to identify carriers of the disease, but also to revise the classification of the disease. Classifications used to be largely organ-based, but are now moving towards a molecular basis in combination with medical imaging. The biomarker must be measured with reliability, precision and reproducibility. The validated method is then generalized in a specific context of prospective research or clinical practice. Any radiology examination that requires measurement and evaluation can be considered a diagnostic biomarker.

The monitoring biomarker or exposure biomarker is measured routinely to assess the status of a disease or exposure to a medical or environmental product. Exposure biomarkers are very important in assessing risk and ensuring the safety of participants during clinical studies by routinely monitoring liver toxicity and cardiotoxicity of drugs. They are also used to measure pharmacokinetic effects and to detect early therapeutic response and disease complications.

An example of a monitoring biomarker is the fluorodeoxyglucose (F-FDG) positron emission tomography (PET) test. The synthetic glucose analogue associated with a radionuclide accumulates in cells with high glycolytic activity. The routine biomarker, called FDG-PET, allows for localization of pathological foci expressing high metabolism and the following of tumor recurrence after cancer therapy.

Predictive and prognostic biomarkers are used to characterize a disease and then to establish its extension.

A predictive biomarker is used to target specific groups of people for whom a particular drug may be useful. It is correlated with the therapeutic activity of a drug. The diagnostic test associated with the treatment is called the companion test. If a molecular change occurs at a specific site, it is possible to treat it and use the same biomarker to monitor the treatment. The effectiveness of the treatment is optimal for patients with the diagnostic marker. In other patients, however, treatment is unnecessary. To date, predictive biomarkers have been primarily genetic variants. The main field of targeted – stratified – therapies is oncology. This fast-growing area is linked to the development of genomic markers for precision medicine, as in the case of cancer patients who have the herceptin (Her-2) receptor, and therefore will respond to trastuzumab (Herceptin®) treatment.

A prognostic biomarker is used to identify the probability of a clinical event, disease recurrence or disease progression. It reflects the pathophysiological process of the disease, independent of drug treatment.

The pharmacodynamic/response biomarker is used to assess the response to therapy. For many drugs, variation in the pharmacodynamic biomarker induces a change in dose during therapy.

Depending on the technology used, three types of biomarkers encompass the parameters:

– biochemical or histological findings detected from a tissue sample;

– biochemicals detected in blood and urine;

– anatomical, functional or molecular findings detected from quantitative imaging.

Medical imaging biomarkers are disease- or tumor-targeted, unlike biological fluid parameters that integrate whole-body information. They are non-invasive techniques that enable spatial and temporal resolution.

2.2.2. *Biomarkers in quantitative imaging*

The term biomarker is applied in all detection methods used in the life sciences. It can be defined as a detectable parameter, whether biochemical, genetic, histological, anatomical, physical, functional or metabolic. By extension, an imaging biomarker is an anatomical, physiological, biochemical or molecular parameter detected by an imaging method used to screen for or assess disease progression.

A biomarker in quantitative imaging is defined as a characteristic that is objectively measured and evaluated as an indicator of normal or pathological biological processes or pharmacological responses to a therapeutic intervention. Techniques such as MRI, CT, X-ray, ultrasound, PET and optical imaging can be used to identify quantitative imaging markers.

Several research projects have image collections available to researchers. Let us take the examples of Alzheimer's disease and lung cancer.

2.2.2.1. *Brain neuroimaging for Alzheimer's disease*

The Alzheimer's Disease Neuroimaging Initiative (ADNI) is a longitudinal multicenter study aimed at developing clinical, imaging, genetic and biochemical biomarkers for the early detection and monitoring of Alzheimer's disease. This study helps in better understanding the disease and its mechanisms and thus the various therapeutic approaches to be tested.

The ADNI began in 2004 under the leadership of Dr. Michael W. Weiner as a public–private partnership with about 20 companies that invested $27 million. At the same time, the NIH Foundation supported the project with $40 million alongside the National Institute on Aging (NIA). The first study, ADNI-1, which was to last five years, was extended to 2011 and then renewed in ADNI-2 and ADNI-3. New participants were recruited from across North America at each new stage, during which clinical and imaging examinations were performed to monitor disease progression. Results are centralized in the image and data archive (IDA) hosted by the University of Southern California Laboratory of Neuroimaging (LINO) [IDA 20]. This platform, accessible to researchers, provides image processing tools for de-identification, visualization and data sharing. The collection gathers more than 80 studies on age-related diseases, among others, integrating more than 53,000 participants aged from 1 to 89 years. Imaging data are complemented by clinical and biochemical information from biological samples.

ADNI-3 combined PET molecular imaging of tubulin-associated unit (Tau) protein with clinical and cognitive assessments, such as PET amyloid detection, MRI and cerebrospinal fluid analysis [WEI 17]. Previous studies have suggested a correlation between Tau protein and cognitive abilities.

MRI in the presence of the amyloid tracer reagent the Pittsburg compound, denoted PiB ([^{11}C]), was used to classify patients according to the Braak scale. A high concentration of PiB and a smaller hippocampus are associated with a high Braak stage [SCH 16]. A strong correlation exists between the memory test and PiB retention at Braak scale levels I/II; and between global cognitive ability and Braak scale levels III/IV and V/VI. The measurement of cerebral Tau protein (cerebral Tau index (CTI)) shows a correlation with the patient's mental state (mini mental state exam (MMSE)).

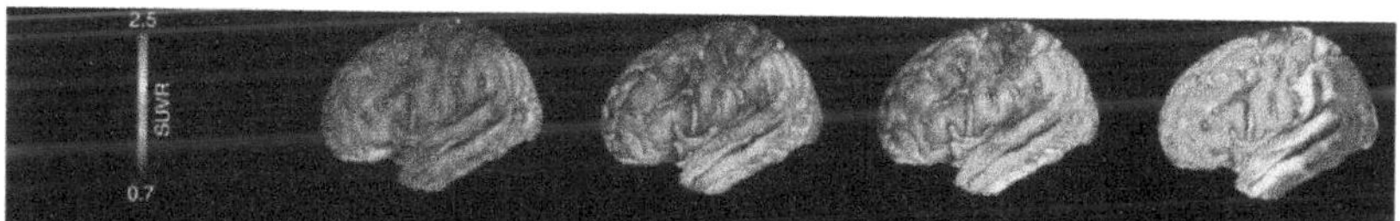

	Braak 0	Braak I/II	Braak III/IV	Braak V/VI
n (YA / OA / AD)	5 / 2 / 0	0 / 25 / 0	0 / 6 / 2	0 / 0 /13
ApoE ε4 (0 / 1 / 2 allele)	2 / 0 / 0 (5 N/A)	18 / 7 / 0	4 / 3 / 0 (1 N/A)	6 / 5 / 2
% *APOE* ε4 +	28	28	38	54
MMSE	28.9 (1.5)	28.8 (1.3)	27.4 (3.2)	18.7 (5.1)
PiB DVR index	1.01 (5 N/A)	1.13	1.35 (1 N/A)	1.80 (2 N/A)
% PiB + (≥1.06)	0 (5 N/A)	44	71 (1 N/A)	100 (2 N/A)
Hippocampal volume (%ICV)	.54	.44	.42	.44

Figure 2.7. *Classification of patients with Alzheimer's disease according to the Braak scale. For a color version of this figure, see www.iste.co.uk/arrighi/biobanks.zip*

COMMENTARY ON FIGURE 2.7. – *Five healthy young adults (YA), 33 healthy older adults (OA) and 15 Alzheimer's disease patients (AD) classified according to several parameters: the proportion of the ApoE e4 and ApoEe4+ allele; mental status (MMSE); PiB retention percentage; and hippocampal volume as a percentage of intracranial volume (ICV). PiB DVR: distribution volume ratio (DVR) of Pittsburg compound [^{11}C] PiB. Figure taken from the journal* Neuron, *presenting the imaging of Tau protein by Scholl's team [SCH 16].*

In summary, several biomarkers change with the progression of Alzheimer's disease:

– beta amyloid measured in cerebrospinal fluid or by PET molecular imaging;

– Tau protein measured in cerebrospinal fluid assesses the degeneration of neurons;

– cerebral atrophy, mainly located in the temporal lobe, measured by structural MRI;

– cognitive tests assess memory loss and general cognitive decline.

2.2.2.2. *The Cancer Imaging Program, a lung cancer screening data consortium*

The Cancer Imaging Program at the National Cancer Institute supports in vivo medical imaging in cancer research. It has assembled an extensive cancer imaging archive accessible for research that combines imaging with genomic information from collections.

The imaging program provides standardized models and validated methods to optimize imaging data [CAN 20]. It is divided into several branches: molecular imaging for the development of preclinical and diagnostic tools, imaging technology that promotes translational research from medical instrumentation to cancer diagnosis and therapy and, finally, clinical trials to study the effectiveness of imaging techniques in diagnosis.

The US National Lung Screening Trial was launched in the early 2000s to show that modern imaging could reduce lung cancer mortality by detecting it early. More than 50,000 participants aged 50 years and older who had smoked for more than 20 years were examined and followed for 10 years with chest CT and conventional radiography [NAT 11]. Unfortunately, the criteria for reading the CT images were not sufficiently selective, and the expected benefit was not demonstrated.

Usually, detection of pulmonary nodules is done from high-resolution CT images. Radiologists manually outline each nodule, which is very time-consuming and tedious. Automatic detection is a tool in development that facilitates the work of radiologists and improves the reliability of the diagnosis. One approach is to submit a collection of images to a panel of experienced radiologists and collect the results, which are then compared with the new algorithms.

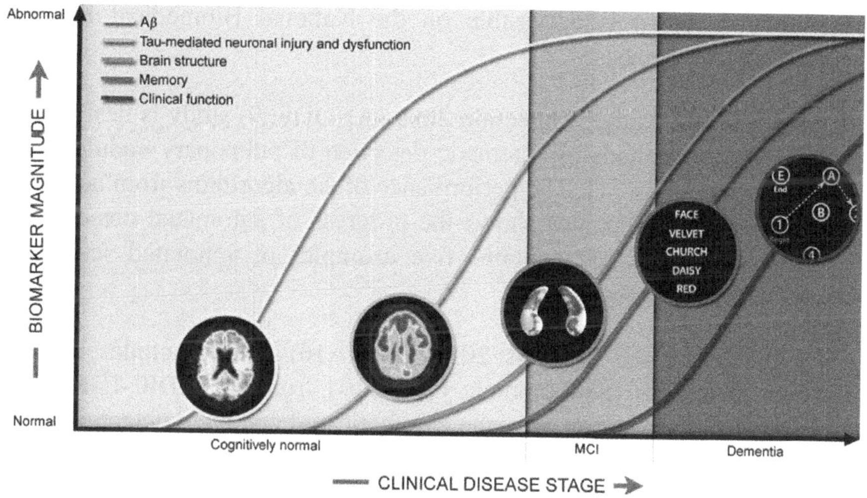

Figure 2.8. *Alzheimer's disease marker profile from normal cognitive state to dementia. Imaging biomarkers detect early brain damage by locating areas of amyloid beta (Aβ) concentration, Tau protein and affected brain structures such as the hippocampus. Mild cognitive impairment (MCI) and then severe cognitive impairment appear at an advanced stage of the disease. Figure taken from the ADNI website [AND 20]. For a color version of this figure, see www.iste.co.uk/arrighi/biobanks.zip*

There are three open access data files for all researchers: the LIDC-IDR, ANODE09 and LUNA16 collections of chest scans. These thoracic scans are used to develop automated diagnostic methods for the detection of pulmonary nodules and their classification. We examine below the objectives of these three studies and the access to their databases.

The Lung Image Database Consortium (LIDC) and the Image Database Resource Initiative (IDRI) comprise a large collection initiated by the NCI and the NIH. The LIDC-IDRI[1] collection includes 1,018 thoracic CT scans from academic and private research centers [ARM 11]. Each scan is independently annotated by four radiologists. Manually labeled nodules are classified into three categories: nodule ≥ 3 mm, nodule <3 mm and non-nodule ≥ 3 mm for a pulmonary abnormality. For nodules larger than 3 mm, annotations describe nodule diameter, calcification, internal structure, texture, presence of lobules and spicules [MCN 07]. Part of the collection

1 https://imaging.cancer.gov/informatics/lidc_idri.htm.

(84 annotated scans) is accessible on the National Biomedical Imaging Archive (NBIA[2]) database.

The Automatic Nodule Detection 2009 (ANODE09) study is designed to compare several methods of automatic detection of pulmonary nodules from chest CT scans [VAN 11]. The performance of six algorithms from academic groups or private companies shows the progress of automated detection of CT scans. The collection contains five examples of annotated scans and 50 scans without annotation[3].

The Lung Nodule Analysis 2016 (LUNA16) study includes a large collection of images (888 thoracic CT scans) from the LDIC-IDRI. The objectives of this study are, first, the automatic and complete detection of all pulmonary nodules and, second, to reduce the number of false positives [SET 17]. The originality of this study is to have combined several algorithms, which has resulted in a sensitivity higher than 92% and less than one false positive per scan. The system has detected nodules that were not previously marked by the experts in the LIDC-IDRI study.

In 2017, the annual Data Science Bowl competition was hosted by the US National Cancer Institute and Kaggle. The question posed to the 1,972 participating teams was, "Can you improve lung cancer detection?" Participants analyzed more than 1,000 high-resolution scans from the National Cancer Institute.

Two researchers from Beijing's Tsinghua University, Liao and Li, have been awarded for their model for assessing the malignancy of lung nodules with deep learning. They proposed a two-step algorithmic model: detection of suspicious lung nodules, and then selection of the five major nodules evaluated for their probability of being cancerous [LIA 19].

2.2.3. *Artificial intelligence for automatic reading of medical images*

Medical imaging and artificial intelligence have transformed the practice of radiology. In 2019, the *Bulletin of the National Academy of Medicine* confirmed the emergence of artificial intelligence in the medical field [BRU

2 imaging.nci.nih.gov/ncia.

3 https://anode09.grand-challenge.org/.

19]. In this general journal, Brunelle et al. reviewed the advances in medical image digitization and algorithm development through deep learning that allow for automatic image processing and cancer screening with higher reliability than radiologists. According to the study by Bruno et al., retrospective analyses of imaging examinations showed an error rate of 30% for all medical imaging techniques [BRU 15].

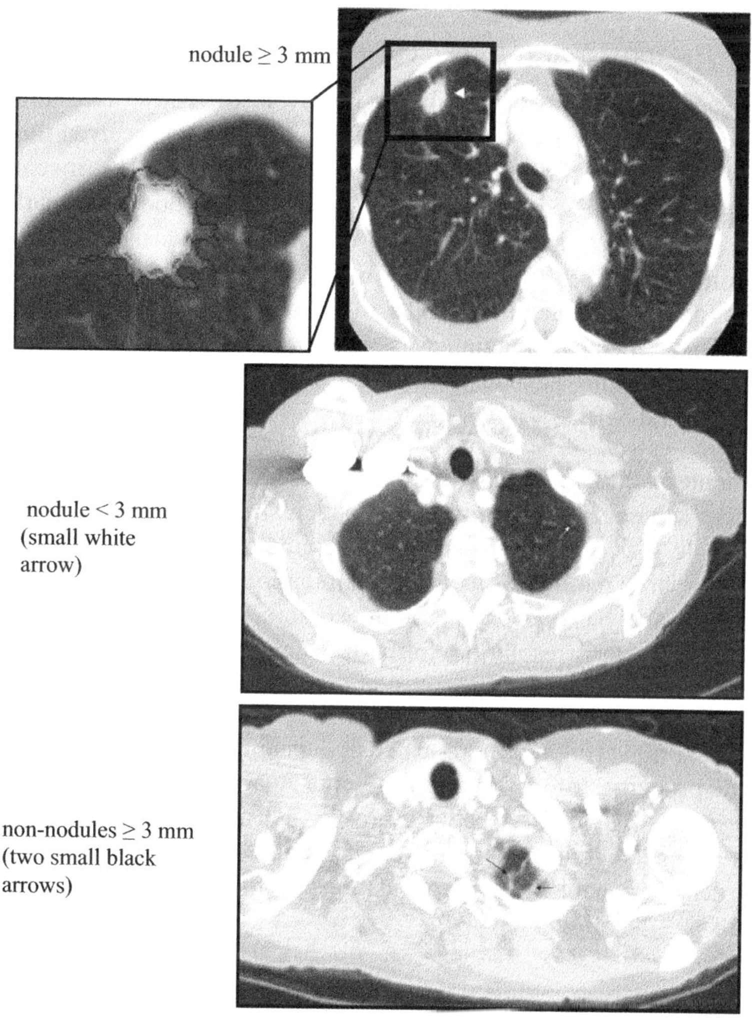

Figure 2.9. *Examples of lung lesions defined according to the LIDC-IDRI study in three categories: nodule larger than 3 mm (nodule ≥ 3 mm), nodule smaller than 3 mm (nodule < 3 mm) and non-nodule lesion larger than 3 mm (non-nodule ≥ 3 mm). For a color version of this figure, see www.iste.co.uk/arrighi/biobanks.zip*

COMMENTARY ON FIGURE 2.9.– *The top insert shows a nodule annotated by three radiologists in different ways (blue, pink and red outlines). Figure adapted from the results presented in the study by McNitt-Gray et al. [MCN 07].*

First, current bio-imaging techniques are presented, then the image processing steps leading to automated image reading are described.

2.2.3.1. *The most frequent medical imaging techniques*

Medical imaging was born with the discovery of X-rays used in the first radiographs in 1895. German physicist Wilhem Röntgen received the Nobel Prize in Physics in 1901 for his discovery of X-rays. The invention of radiology revolutionized the exploration of the human body and marked the beginning of medical instrumentation. With this non-invasive technique, the restitution of images of the internal structures of the human body gave the clinician access to anatomical and morphological characteristics of the organs. Then, medical imaging was perfected with the entry of artificial intelligence, bringing several techniques of clinical image acquisition such as CT, PET and MRI.

Computed tomography, also called a CT scan, is a common examination for cancer screening and follow-up of patients. This technique is based on the principle of X-ray absorption in the volume of tissues and organs. It has generated a large number of imaging examinations with high-resolution data available for analysis. Depending on the equipment, CT images have various technical parameters (resolution, format), which are sometimes incompatible. After collection and before analysis, a pre-processing step is essential.

MRI is a powerful and more expensive technique than CT. Based on the magnetic properties of matter, it gives better image quality if the magnet has a high intensity. MRI visualizes proton density with several types of contrast. The combination of different excitation sequences (T1, T2, or proton density) enables the identification of different types of tissue and structure.

Positron emission tomography, also called positron emission tomoscintigraphy or a PET scan, is the least commonly used technique because its resolution is lower than that of CT. This examination is based on the injection of a weakly radioactive tracer into the blood. For example, radioactive glucose is metabolized more rapidly in tumor cells and provides functional information on the metabolism and proliferation of the tumor.

2.2.3.2. *Segmentation of the image and marking of the region of interest*

Medical images are often annotated by radiologists to specify the location of a lesion or tumor. Segmentation, also called delineation, is the act of selecting a portion of the image for further analysis. In practice, precise delineation of target volumes using specialized software is performed by the radiation oncologist with a CT scan or other anatomical (MRI) or metabolic (PET scan) imaging techniques.

Image segmentation is the process of dividing the image into several parts, each corresponding to a collection of pixels in 3D, called volumetric pixels or voxels. The voxel stores colorimetric information, spatial coordinates, the size of the unit used and the texture, and it requires large storage capacities.

The purpose of segmentation is to provide quantitative measurements such as the volume or thickness of a structure. The automation of segmentation is expected to reduce the burden of this tedious and error-prone task. Many methods have been developed, from graphical tools and machine learning algorithms to, more recently, deep learning by convolutional neural networks (CNNs).

The manually or automatically delimited areas can correspond to pathological lesions of the organs or identify a region or volume of interest (ROI, VOI).

They provide the analysis with quantitative measures such as:

– measurement of tumor volume;

– quantification of metabolic activity in the ROI;

– quantification of image features (texture and shape) from radiomics properties.

The performance of the automatic segmentation is tested on two groups. The supervised group, a subgroup of the dataset, contains the annotated data. The second group, unsupervised, contains the unannotated dataset. The algorithm is trained on a first dataset (training set) and then evaluated on a second dataset not presented during training.

The evaluation compares the segmentation measurements predicted by the software with reference measurements. Ideally, these reference values reflect the significant ground truth. In practice, the reference measures correspond to the manual annotations of the experts, with the known caveat regarding the variability of the annotations between several experts and of the manual rectifications of the same expert. Beyond the global performance of the algorithm, it is important to measure the quality of an individual analysis and detect cases where the segmentation is erroneous. This quality control parameter is crucial in the automatic segmentation of large-scale studies such as clinical screening and population-based studies.

2.2.3.3. *Descriptor extraction and classification*

During the segmentation, the ROI is located manually or automatically. This region is then analyzed by the statistical algorithm. Radiomic quantification tools allow for "decoding" the phenotype of an ROI. After analyzing CT scans of 1,019 lung cancer patients, Aerts et al. presented 440 descriptors extracted from the imaging data. In the results published in *Nature Communications* they concluded that three categories of descriptors globally quantify the ROI: image shape, texture and intensity [AER 14].

Routinely, the evaluation of tumor response to therapy is measured by the Response Evaluation Criteria in Solid Tumors (RECIST), revised in 2009 [EIS 09]. This simple method is based on the measurement of the tumor and the number of measurable lesions [JAF 06]. For a tumor, the diameter is measured along the longest axis, while the measurement follows the shortest axis for the lymph node. The sum of the diameters of the target lesions is followed in the clinical trial to assess tumor response or progression. This morphological measurement is one-dimensional and does not provide information on anatomical volume. Therefore, descriptions complete the data by specifying the global appearance of the lesion – heterogeneous or spiculated – while remaining qualitative and therefore imprecise. In three dimensions, the analysis of the shape integrates the measurements of surface, volume and contour.

The intensity of the signal depends on the physical parameters of each tissue. Each digital image records the value of the signal at each point in space. The ROI is divided into small surface elements (pixels) or volume elements (voxels). The signals measured in each pixel are represented on a gray level histogram, giving the signal intensity distribution. The spatial distribution matrix of the pixels extracts the mathematical descriptors of the heterogeneity of the intra-tumor texture.

Among the large number of descriptors obtained for each patient are shape, signal intensity distribution and texture data. Only relevant data are retained for classification and then stratification of patients into groups of common phenotypes. Robust and independent descriptors are extracted. The reduction of the number of descriptors is based on the elimination of features:

– non-reproducible: intra-observer reproducibility is validated by the intra-class correlation coefficient applied by the same radiologist. Inter-observer reproducibility is confirmed by the concordance correlation coefficient applied by several different radiologists;

– redundant: a correlation matrix evaluates the redundancy between descriptors. Correlated descriptors are deleted.

The review by Vial et al. presents the crucial role of imaging descriptor extraction in cancer detection. To facilitate detection, machine learning and deep learning methods are being developed with the goal of collecting more and more information on all patients and tumor types [VIA 18].

2.2.3.4. *More accurate diagnosis of medical images with artificial intelligence neural networks*

Scientists spend the majority of their time preparing data rather than exploring it. The Forbes survey institute estimated in 2016 that data scientists spend 60% of their time preparing and organizing data [PRE 16]. During this step, also called data curation, the quality of the data is checked. The second activity of the experts is data collection, occupying 20% of their time. The remaining time is used to plan data mining activities (9% of their time) and design learning algorithms (3% of their time). In summary, about 80% of a data scientist's time is spent managing data upstream of analysis –

and that is not the most enjoyable part, according to the 80 data scientists surveyed in the Gil Press survey for Forbes [PRE 16]!

In his article on the future of radiology, Hugo Aerts sees that "the future of radiology is bright, as novel artificial intelligence algorithms have the potential to increase the clinical efficacy and quality of radiology, while making it faster and cheaper" [AER 18]. Indeed, artificial intelligence enables quantitative assessment of radiographic features. Machine learning is broadly defined as programming methods based on experimental data. These algorithms are able to learn from the data and improve autonomously.

Deep learning, a subcategory of machine learning, is the most developed form of artificial intelligence in imaging. The technology is based on the concept of CNNs and uses both supervised and unsupervised human learning methods. In practice, two sets of patient metadata are processed in two different schemes. In the first dataset, deep learning models are developed and evaluated. Network validation is then performed on the second dataset. After a "training" period on known data (training dataset), the algorithm performs diagnostic tasks, the most common of which are detection, object segmentation, automated extraction of imaging descriptors and identification of nonlinear relationships in complex data.

Aerts' work has shown that machine learning techniques such as neural networks have the ability to describe tumor phenotypes with greater predictive power than routine clinical measurements. Recent advances in deep learning have shown successful applications in image analysis. Unfortunately, these methods require a large number of annotated examples to train a model. Furthermore, the volume of data required by deep learning exceeds the storage capabilities of our current systems. Numerous programs are proliferating in both university laboratories and private companies.

In 2017, French company Therapixel trained a deep convolutional network on 640,000 mammograms in a global challenge. In two weeks, the 1,200 participants had to come up with proposals to improve the analysis of mammograms and thus the early detection of breast cancers. Therapixel won the first prize in the Digital Mammography Dream Challenge by distinguishing suspicious mammograms from normal ones better than all competitors: the software reduced the number of false positives by 5% compared to other systems [SCH 20]. Therapixel's next goal is to improve the accuracy of the detection tool and then clinically validate the algorithm in order to bring the software to market.

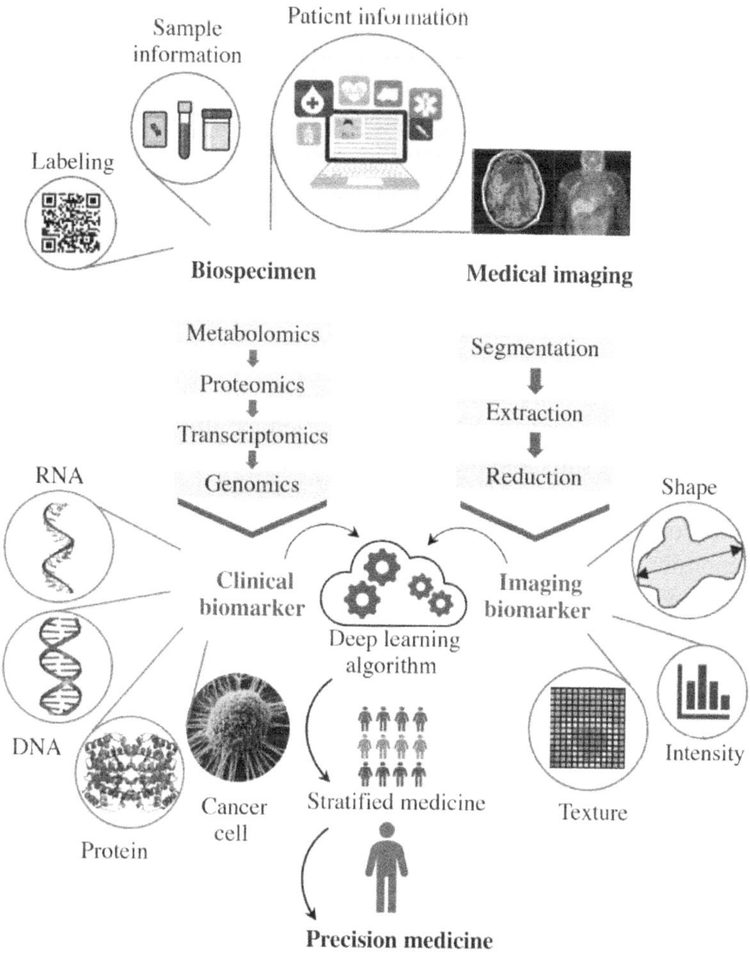

Figure 2.10. *Precision medicine is driving the strategic directions of modern biobanks from sample management to health data management. For a color version of this figure, see www.iste.co.uk/arrighi/biobanks.zip*

COMMENTARY ON FIGURE 2.10.– *Information from biological samples crossed with medical imaging data has enabled medicine to leap forward with statistical algorithms. A massive volume of data is being hosted on a large scale by IT giants who are very interested in the qualified, clear, annotated medical data required for artificial intelligence applied to health.*

Therapixel has opened an image exchange platform resulting from the Radvise-X project. The Radvise portal aims to provide radiologists with a reliable interpretation tool for thoracic scans based on artificial intelligence algorithms, enabling them to save precious time in diagnosis, and even to refine the reading grid in order to reduce the number of cases of overdiagnosis and the resulting additional costs. "Our main objective is to help the first reading," explains Pierre Fillard, co-founder of Therapixel, in the columns of the business newspaper *Les Echos* from March 13, 2019 [DEM 19]. In other words, it is to assist the radiologist without replacing them.

The development of algorithms capable of extracting quantitative descriptors in large numbers has enhanced medical imaging both in terms of the quality of the results and the volume of images processed.

A new concept was born from the correlation of radiomics and genomics: radiogenomics. In concrete terms, medical images reflect genetic and molecular processes.

Using imaging and genomic data (The Cancer Imaging Archive (TCIA) and The Cancer Genome Archive (TCGA)), Zhu et al. investigated the association between radiomic phenotypes of 91 breast cancer carcinomas and genomic features of molecular regulation [ZHU 15]. The team demonstrated that radiomic phenotypes such as tumor size, shape, morphology and texture were closely correlated with gene regulations, miRNA expression and somatic gene mutations.

The data from medical imaging biobanks represent an organized source of information from routine non-invasive medical examinations. The quality of the image acquisition and its cross-use with omics studies improves medical decision-making. Harmonization and quantification of medical images are the two priorities of the healthcare system. Bio-imaging information is an additional tool reflecting the expression of the patient's phenotype.

2.3. The infallible traceability of biobank data

High-quality biological samples are essential for medical research. Several studies have shown that small deviations in pre-analytical variables lead to analytical errors, especially in omics technologies. Reproducibility of results is only achieved with a high-quality sample collection.

We here explain how biobanks ensure infallible traceability, first through the automated processing of the sample and then through the computerized management of the data related to the sample.

Ensuring the traceability of samples and associated data is carried out in terms of an infrastructure and also in a group of infrastructures. This is called an Expert Center. These biobanks with international expertise are specialized and apply the principles of FAIR management of health data, developed at the end of this section.

2.3.1. *Automated sample flow, essential for population biobanks*

In a biobank, all procedures are standardized. The implementation of the new DIN EN ISO 20387 has strengthened the quality management of biobanks by controlling the technical factors influencing the quality of the sample.

Automation leads to:

– the risk of error being reduced;

– reduction in the number of manual steps;

– the continuous traceability of samples;

– simplification of the sample process.

In a biobank, automation can be generalized apart from for the collection step, which remains manual. For a biological fluid, all pre-analytical steps can be automated. For a biological tissue, certain steps such as aliquoting are not automated, except in the presence of a surgical robot. The automation of sample processing includes the aliquoting step and more complex steps such as the extraction of nucleic acids or the separation of the buffy coat (blood mononuclear cells).

Nucleic acid extraction involves several methods, such as magnetic beads and RNA precipitation. A comparative study was carried out by Hantzsch's team by comparing the column and magnetic bead protocols and, in parallel, the manual technique and the automatic technique [HAN 14]. Several kits were assessed, including PAXgene (Qiagen) and Tempus (Life Technologies) Blood RNA Tubes, according to the following performance indicators:

– the yield of RNA, both quantitatively and qualitatively;

– the duration of the extraction, in total handling time and in individual processing time per sample;

– the final result for the mRNA and microRNA expression.

Hantzsch et al. showed that the extraction yields are comparable between manual and automatic protocols, while the quality is similar. They note a clear improvement in RNA extraction with the new extraction kits, especially for microRNA collections.

Automatic aliquoting requires tubes specially adapted to the Society for Biomolecular Screening (SBS) format. Widely used racks of the same format (SBS) allow for the placing of cryotubes of different volumes (0.5 to 1 mL) for automatic pipetting operations. Each tube has a linear barcode inscribed on the tube and a 2D code on the bottom of the tube, automatically read through the open bottom of the rack. The traceability of the whole chain can be controlled, provided that all modules of the chain are compatible.

Another aliquoting format is the Cryo Bio System (CBS) format. Robotic aliquoting automates all operations from aspiration of the biological fluid (0.3 to 0.5 mL) through a permeable cap to heat sealing of the two ends after filling. The heat-sealing process ensures that the vials are watertight and they are usually stored in liquid nitrogen.

The freezing step plays an important role in maintaining the sample's integrity. Ideally, freezing should immediately follow aliquoting, or better yet, both actions should be accomplished simultaneously by pipetting the sample directly into the storage cryotube placed at –20°C. The temperature decrease is the essential parameter of the automated freezing system. A speed of –1°C/min can be controlled manually. However, there are automatic systems that optimize the temperature decrease according to the starting temperature of the sample. Once the freezing temperature is reached, the samples are transferred to freezers or liquid nitrogen containers with very low temperatures. Automatic control of the cryopreservation steps increases the viability rate of the cells. Buhl et al. studied the cryopreservation of blood mononuclear cells activated in T cells in immunotherapy trials. The amount of cells recovered and the viability, phenotype and functional properties of the cryopreserved cells were analyzed [BUH 14]. The automatic freezing system improved the yields of mature and immature dentritic cells. The values obtained were comparable to those obtained from freshly prepared cells. The yield was 50% higher than the standard freezing vessel protocol in the presence of isopropyl alcohol.

An automated storage system at –80°C reduces temperature variations that would be unavoidable without an automated system. Opening the freezer to deposit or remove a sample inevitably induces a temperature fluctuation. However, storage at –80°C presents difficulties because mechanical and electronic systems are not reliable at this temperature.

The automated storage system is composed of several chambers (or zones). In the –20°C zone, robotic arms move to deposit and retrieve the samples. Keeping the mechanical parts at a temperature above –20°C ensures the longevity of the material. The samples are handled in a –80°C zone, without successive passages at different temperatures, thus limiting the risk of deteriorating the sample integrity.

Automatic sample processing is essential in population-based biobanks. To follow a large number of participants and to organize the logistics of the French Constances Cohort, Henny et al. published a guide for the constitution of a biobank associated with epidemiological studies with the general population [HEN 13]. The authors analyze the functions of the biobank and recommend robotic technical solutions. From sample receipt to sample shipment, they describe the sample process in the biobank as automated steps driven by what they term "biobank engineering".

2.3.2. *Computerized management of clinicobiological annotations*

In biobanks, the biobank information management system (BIMS) oversees all the data in the laboratory information management system (LIMS). The BIMS ensures the traceability of the samples and also of the computerized systems storing patient information (electronic medical record) and of the sample storage conditions (temperature control of freezers).

Figure 2.11. *The sample life cycle, driven by the BIMS*

COMMENTARY ON FIGURE 2.11.– *The workflow provides patient information and sample data. The biobank stores all of this data in the BIMS software, which provides sample traceability and centralizes all of the data associated with the sample. The figure is adapted from the publication by Le Quéau et al. with their kind permission [LEQ 20].*

The BIMS follows the biobank workflow and provides the scientific and clinical data needed for medical research. Researchers and clinicians are increasing their efforts to decipher the metabolic processes that trigger disease. Omics technologies – genomics, transcriptomics and proteomics – have generated massive data. Analyzing these data and creating data platforms is an ongoing challenge. This central role of data and information collector is played by the BIMS.

2.3.2.1. *The complex architecture of the BIMS*

The sample process in the biobank is supervised by the LIMS software at each step: from reception to preparation and storage and, finally, distribution of the sample or its associated data. This software facilitates the standardization of procedures and ensures process control. Unlike the LIMS, the BIMS integrates both the characteristic information for the sample and the clinical data related to the patient, such as informed consent. The software is hosted by the local area network (LAN) connecting the computers included in the predefined area. Several databases are integrated in the BIMS. Their information can be extracted and made available to researchers via a virtual catalog.

The hospital information system (HIS) and the clinical databases (CDBs) are the source of clinical information on patients, which is governed by strict regulations, especially with regard to patient anonymity [LÉG 78]. These databases are accessible on the hospital's servers in the case of a hospital biobank or they can be connected to by the BIMS through a specific computer language [SPA 11]. In both cases, they comply with the European regulations on personal data protection [RGP 16].

The BIMS can also integrate temperature control devices, for example, freezers or the monitoring of liquid nitrogen levels in containers. These data are stored separately in the quality management system (QMS). However, in case of non-conformity, the procedures for initiating corrective actions can be accessed and documented in the BIMS. Finally, the BIMS can provide interoperability with analytical instruments such as high-throughput sequencing machines, eliminating the need for manual data entry into the software.

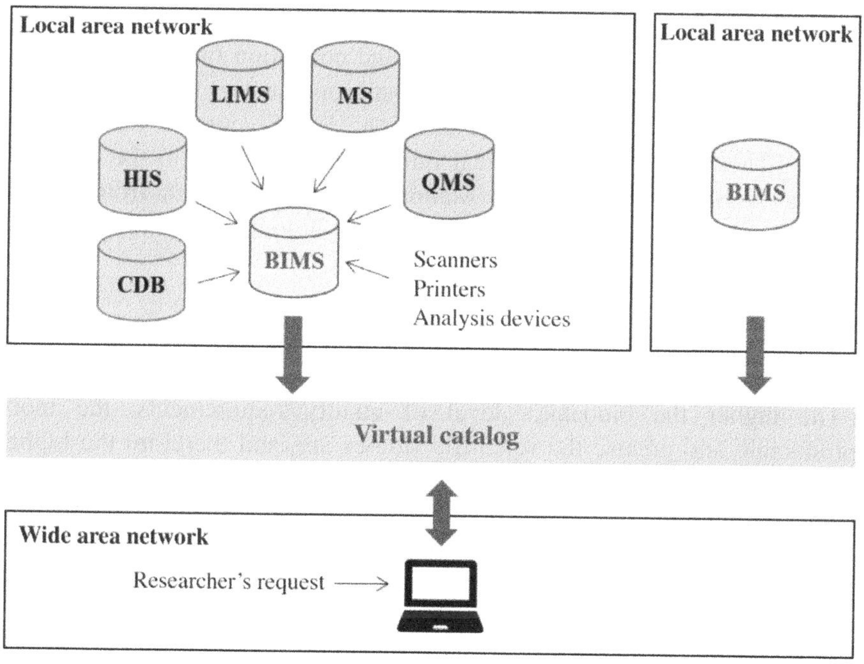

Figure 2.12. *Databases connected to the BIMS. CDB: clinical database; HIS: hospital information system; LIMS: laboratory management information system; MS: monitoring system; QMS: quality management system; BIMS: biobank information management system*

COMMENTARY ON FIGURE 2.12.– *The software is hosted by a local area network that connects multiple computers and analytical devices. The CDB and the HIS are connected to the BIMS, as well as the QMS and the LIMS. The figure is adapted from the publication by Le Quéau et al. with their kind permission [LEQ 20].*

The BIMS centralizes sensitive and confidential information shared over the local network connecting the biobank, hospital or corporate terminals. However, researchers may want to locate a particular sample or collection. To answer their request, a virtual catalog is available via a wide area network (WAN). This online catalog aggregates information from a single site or from several sites of a biobank, such as in the UK Biobank.

The BIMS opens up a wide range of actions to an unlimited number of users in mono- or multi-site contexts. As the information comes from a hospital, an internal/external laboratory or a biobank, access to the BIMS is configured according to the user profile. Controlled access allows identified members to enter specific information. At the time of collection, the physician assigns the barcode and enters the collection time. The clinician adds the patient information and the biobank operator oversees the sample process. All access requests and uses of the BIMS must be discussed in advance. The biological and clinical information collected in the BIMS is varied, as is the format of the data. Manually implemented data from various sources is time-consuming, whereas the BIMS automatically integrates the data and reduces the number of data entry errors.

2.3.2.2. *BIMS software, centralized management of samples and their associated data*

The higher the biobank's level of quality requirements, the more reproducible and precise the scientific studies are, and therefore the higher the quality. According to the comparative study conducted by Orchard-Webb on the different computerized systems dedicated to biobanks, the BIMS has three major principles [ORC 18]:

– management of a wide range of samples such as blood, tissue, saliva and DNA;

– integration of electronic medical records;

– obligation to comply with the regulations.

In concrete terms, the needs are varied and must be managed in real time at several levels. In the laboratory, sample monitoring during the aliquoting and analysis preparation phases ensures sample integrity. Upstream, logistics management and cold chain control are critical. Then, at the end of the cycle, information storage needs, especially with several sites, must be anticipated.

Companies such as Technidata, Modul-Bio, Thermo Fisher Scientific, LabVantage Solutions Inc., Genohm and Brooks Life Sciences have developed BIMS software to meet the specific demands of biobank users. There are differences in interoperability, module customization and maintenance between the software packages. The cost includes installation, maintenance and updating of the software. It also depends on the number of sites, licenses and modules required. The use of a BIMS is not mandatory, nor is it compulsory to purchase a pre-configured BIMS from commercial vendors. The alternative is to program one's own software (made in-house) from free software (open source), which requires a solid background in computing.

The measure of a biobank's quality is the number of high-value samples and the chain of custody surrounding these valuable samples. Data quality is at the heart of a biobank's business because samples are only valuable if they are associated with quality clinical and biological data. Although a BIMS is not mandatory for accreditation with the NFS 96-800 standard for biobank quality, it facilitates the management of large-scale collections and improves data consistency for certification purposes. In a BIMS, laboratory and sample data are stored in an electronic system that improves the robustness of quality systems.

The BIMS management system is strongly recommended to improve and maintain the quality of scientific results. Upstream of clinical trials, clinicobiological annotations such as phenotypic, genotypic and environmental information collected in the BIMS can be used to identify patient groups. For example, the analysis of patient DNA sequences determines different patient groups in order to set the appropriate drug concentration for the processing of each group [OLU 07]. The integration of as much patient and sample-related data as possible guides clinicians and researchers in the discovery of new therapeutic strategies.

Data may come from many different studies, formats may vary and data collection methods may vary. BIMS data management provides a single access point, improving the availability of clinical and biological annotations. On a daily basis, hospitals produce vast amounts of data, but unfortunately it is not used to its full potential. Use of a BIMS and data centers are the solution to collect all the data needed for studies in silico.

Large pharmaceutical companies are interested in building their own biobanks to develop research for new drugs and in vitro diagnostic tests [BEG 12]. Smaller companies can collaborate with public biobanks to gain access to rare samples. Building a private collection is expensive and not cost-effective if it is only used in one project. Instead of collecting FFPE blocks from patients, a pharmaceutical company can collaborate with a biobank to obtain patients' histological slides and their associated data. Finally, both parties exchange their mutual expertise.

During drug development, a large number of samples are processed, analyzed and stored. A partnership with a biobank brings expertise in sample collection and storage. Large pharmaceutical groups target very high-quality biobanks, which emphasize sample traceability and provide all the clinical and biological information; in short, all the actions gathered in the BIMS.

2.3.3. *International Biobanks of Excellence or Expert Centers*

Future innovations in healthcare will only be possible if human samples are of high quality and associated with clinicobiological information. As these resources are limited and their ethical and legal framework is quite strict, their use must be optimized. Conducted under standardized conditions and on an international scale, sample analyses must allow for:

– access to and sharing of analytical data;

– storage of information from the samples so that they can be used for future product development.

The BBMRI-ERIC Biobanks of Excellence or Expert Centers fulfill both missions. These centers are non-profit organizations based on a public–private partnership model. They carry out the analysis of samples in the country of origin in accordance with standardized and international protocols. The generated data, called primary data, are shared via the BBMRI-ERIC biobank network. These Expert Centers centralize the results of public and private research, combining academic and industrial expertise. This collaborative research ensures compliance with the legal and ethical framework for all partners.

The main characteristic of a high-quality Expert Center is broad medical and scientific expertise. Such centers have access to state-of-the-art technology, a high level of standardization and professional quality management.

A major obstacle for multicenter studies is that the analytical platforms in omics technologies and the upstream pre-analytical processes are not sufficiently standardized to integrate the analytical data in different centers. A high level of standardization can be achieved by building a network of interconnected Expert Centers implementing harmonized standard operating procedures (SOPs), that is, common certification and accreditation procedures. The data generated by such standardized platforms, on an international scale, facilitate the integration of analyses aiming to answer medical and biological questions. Finally, biological materials can be used more efficiently by providing added value that is shared with the entire scientific community.

Specialized centers can afford technology maintenance that is unaffordable for smaller, non-specialized facilities. These analytical advances coincide with the general trend of outsourcing specific tasks to specialized providers. For example, new sequencing technologies have an enormous analytical capacity that cannot be used and managed properly by each institution. Therefore, Expert Centers can operate at lower costs than academic or industrial research laboratories.

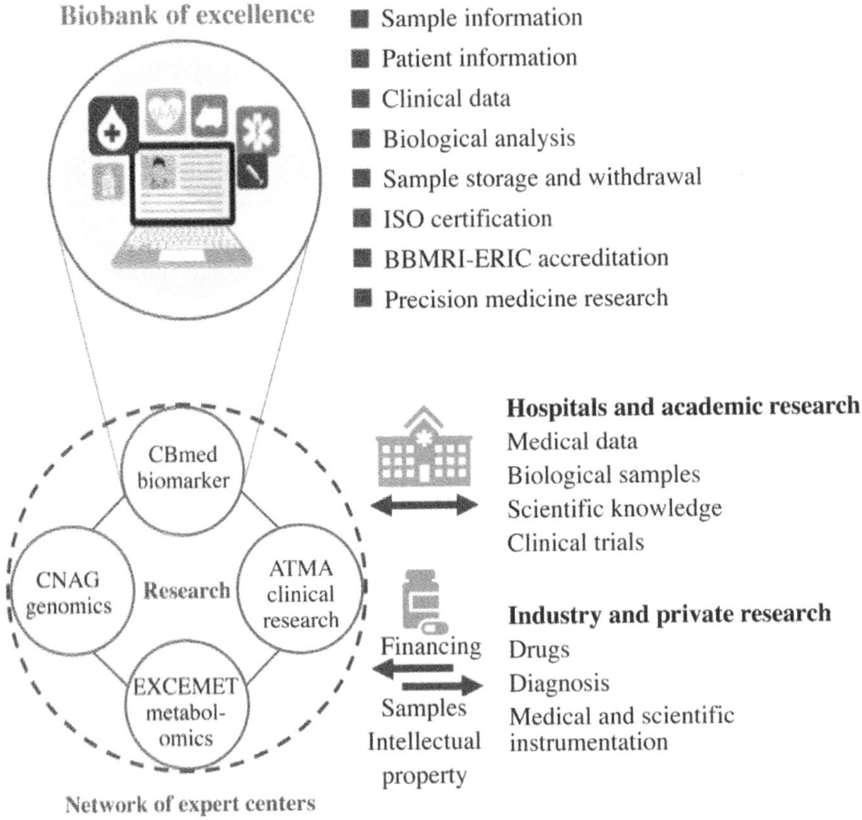

Figure 2.13. *BBMRI-ERIC Biobanks of Excellence*

COMMENTARY ON FIGURE 2.13.– *The network of Expert Centers links Biobanks of Excellence from different fields and countries: the Center for Biomarker Research in Medicine, CBmed (Austria); the platform for molecular analysis of clinical tissues, ATMA (Italy); the National Genomics Center of Barcelona, CNAG-CRG (Spain); and the Expert Center for Metabolomics, EXCEMET. The Expert Centers interact with academic centers of public research and private companies involved in medical and pharmaceutical research.*

The provision of biospecimens and their clinicobiological analysis are not considered inventions. Biobanks and Expert Centers do not warrant specific intellectual property. However, in the context of a collaboration, joint intellectual property is adopted between industrial partners and Expert

Centers. The general policy is to anticipate the agreement of the parties by discussing upstream the sharing of rights so as to minimize the obstacles during the exploitation of the discovery. In the case of a joint invention, the academic partners should receive royalties recognizing the contribution of public resources in expertise and personnel, while the industrial partners extend the intellectual property to the final product. In the absence of an industrial partner, the objective of expertise biobanks is to translate the results into industrial or medical applications by licensing the discovery or by creating start-up companies.

Biobanking Expert Centers are a resource for the pharmaceutical and biological industries. The work of van Ommen et al. compiles their missions and characteristics [VAN 15]:

– participation of research leaders implementing cutting-edge technologies;

– use of standards and reference samples;

– participation in test evaluations and circular analyses;

– publication of SOPs for pre-analytical steps, molecular analysis and data management;

– establishing confidentiality and intellectual property procedures;

– compliance with legal and ethical regulations;

– commitment to effective contract and project management;

– certification (ISO);

– BBMRI-ERIC accreditation through periodic external audits.

One of an Expert Center's first activities after defining the scope of work and the standard procedures is to obtain reference samples and participate in the inter-biobank testing program. These regular testing programs are performance evaluation tests of the participating biobanks. This evaluation serves to improve future interoperability and standardization.

The following biobanking Expert Centers have various specialties:

– the CBmed Center for Biomarker Research in Medicine;

– the ATMA clinical tissue molecular analysis platform;

– the National Genomics Center of Barcelona;

– the EXCEMET Expert Center for Metabolomics.

2.3.3.1. *The first expert biobank, the Center for Biomarker Research in Medicine in Austria*

The Center for Biomarker Research in Medicine (CBmed) combines an excellent research infrastructure, scientific expertise, medical knowledge, and national and international industry partners for medical research on biomarkers in medicine. Established in 2016 in Graz, Austria, CBmed was the first Expert Center in the BBMRI-ERIC network. This public research center is owned by the Medical Universities of Graz (43.5%) and Vienna (20%), the Technical University of Graz, the Joanneum Research Institute and the Austrian Institute of Technology. The consortium has 34 industrial and 23 scientific partners.

The term "stratification" describes the characterization of a disease. In breast cancer, each subtype of the disease requires a different treatment. The more accurately the subtype is characterized, the more targeted the therapy can be. There are hundreds of biobanks that share their data with researchers and clinicians in an effort to identify these types. For 30 years, the Medical University of Graz has been collecting high-quality biological samples. With CBmed, the goal is to develop a methodical approach platform for biomolecular research in collaboration with all European biobanks.

The commitment of the BBMRI-ERIC biobanking Expert Centers represents a clear commitment to the introduction of quality standards in the analysis of human samples. "Through the introduction of quality criteria and the standardization of procedures, the data obtained are made internationally comparable – which means an acceleration of research and thus better treatments", emphasizes Professor Jan-Eric Litton, managing director of BBMRI-ERIC [BBM 16].

By identifying new biomarkers, it will be possible to treat cancer and metabolic and inflammatory diseases individually for each patient. "Over the last few years, the knowledge that each person responds differently to each treatment has become more widespread. It is the long-term goal of medical research to make personalized, tailor-made treatment available to each individual patient", explains Professor Thomas Pieber, scientific director of CBmed [MED 16].

2.3.3.2. *The ATMA clinical tissue molecular analysis platform in Italy*

In Italy, Archive Tissue Molecular Analysis (ATMA) is designed to improve the levels of standardization of molecular analysis and thus accelerate clinical research. It is a non-profit organization based on a public–private partnership. It integrates public and private research and development activities by providing biological samples and data, as well as broad scientific and medical expertise in tissue preservation and molecular analysis. A review by Stanta et al. investigated the role of tumor heterogeneity in therapy resistance and poor reproducibility in clinical cancer research [STA 16]. They demonstrated that there is a need to develop new tools and biomarkers to study tumor heterogeneity. Technically, the study of clinical tissues relies on the control of several factors: pre-analytical conditions, standardization methods, internal and external quality assurance, and characterization of tumor heterogeneity.

Backed by the biotech industry association Assobiotec, the ATMA Expert Center (ATMA-EC) platform is a consortium composed of the following organizations:

– the Oncology Reference Center of Aviano (*Centro du riferimento oncologico* (CRO)) is run by the Italian Cancer Research Institute (*Istituto di Ricovero e Cura a Carattere Scientifico* (IRCCS)). Its mission is to improve public health by advancing medical knowledge and training physicians and scientists in oncology. This non-profit, public organization is equipped with scientific platforms for flow cytometry, molecular biology, biobanks and a clinical research department;

– the Laboratory of Pathology, Clinical Microbiology and Virology of the University of Milan-Bicocca. The University of Milan-Bicocca is a public university funded by the State. In collaboration with Assobiotec, it develops international training programs;

– the Diagnostic and Nuclear Research Institute (SDN IRCCS) of Naples is a platform specialized in radiomics. Two research laboratories support its activity: the Imaging Processing Center (NAPLAB) and the statistics and bioinformatics, radiopharmaceuticals, nanotechnology and biobanking laboratory;

– the biotech company AB Analitica located in Padua is ISO-certified. It designs, develops, produces and markets in vitro diagnostic equipment for molecular biology and laboratory diagnosis of biological analyses;

– Hiantis, located in Milan, develops molecular diagnostic kits to detect human papilloma virus (HPV) infections that develop into cervical cancer.

Analyses are performed in strict compliance with standardized quality procedures and in accordance with ethical, legal and social regulations. The centers verify the reproducibility and validation of new molecular biomarkers and imaging biomarkers, under internationally harmonized standard operating conditions. The goal is to reduce the cost of clinical research, decrease the time required for development and accelerate the clinical application phase for new therapeutic targets and biomarkers.

2.3.3.3. *The National Genomics Center of Barcelona, CNAG-CRG*

The National Genomics Center for Analysis in Barcelona (*Centro nacional de análisis genómico* (CNAG)) was named a BBMRI-ERIC Expert Center in 2018, but it has been in existence since 2009. It was created with the mission of conducting DNA sequencing projects in collaboration with Catalan and Spanish researchers and the international scientific community. It is a non-profit organization funded by the Spanish Ministry of Economy and Competitiveness and the Catalan government. Since 2015, CNAG has been integrated into the Center for Genomic Regulation (*Centre de regulacio genomica* (CRG)). This center is equipped with important sequencing capabilities and a powerful IT infrastructure.

CNAG research projects focus on the development of analytical and bioinformatics tools for next-generation sequencing and analysis of genomic information in many areas, such as cancer genetics, rare diseases, host–pathogen interactions, endangered species preservation, evolutionary genomics studies, 3D genomics, epigenomics and agricultural plant breeding.

The CNAG is ISO 17025:2005 certified for its next-generation sequencing platform. This international standard establishes the general requirements for competence when performing tests and/or calibrations. It demonstrates the high level of expertise and technical competence of the CNAG and follows the ISO 9001:2008 certification specifying the requirements for the quality management system. “Both international standards are strategic tools that minimize errors, increase productivity and ensure CNAG meets the highest standards possible to fulfill the expectations of precision medicine”, says Ivo Glynne Gut, CNAG director [CNA 16].

In addition to these ISO standards, CNAG is referenced by industrial groups specialized in medical diagnostics. The CNAG is the certified service provider for Illumina and Agilent for exome sequencing. The center is also a member of the Roche sequencing program thanks to its expertise with targeted enrichment systems with the SeqCap® EZ kit.

In the article by Auffray et al. entitled "Making sense of big data: Towards an EU action plan" and published in the journal *Genome Medicine* in 2016, the authors showed that genome sequencing and high-resolution imaging technologies are the driving forces in medicine [AUF 16]. They show the generation of an exponential amount of Big Data that we must learn to manage. According to the authors, the exploitation of data from genomic medicine, medical imaging, mobile applications and connected health objects is hampered by historical, technical, legal and political obstacles. European healthcare systems are different and fragmented. The lack of harmonization in data formats, processing and analyses generates incompatibilities. The authors suggest areas for improvement in methods, tools, analysis and data availability. In their view, the creation of a single European health market would improve the health of all Europeans.

2.3.3.4. *The EXCEMET Expert Center for Metabolomics*

The Expert Center for Metabolomics (EXCEMET) is an international consortium of leading institutions in the field of metabolomics [EXC 14]. This non-profit public–private consortium is composed of academic and industrial partners. The founding members are the University of Florence in Italy (Consorzio interuniversitario risonanze magnetiche di metallo proteine), the Medical University of Graz (Austria), the University of Berlin Metanomics Health (Germany), the University of Gothenburg (Sweden) and the UK's University of Birmingham. Other international partners have joined the Expert Center: Helmhotz Center Munich (the German Research Center for Health and Environment), Steno Diabetes Center Gentofte (Denmark), the European Bioinformatics Institute European Molecular Biology Laboratory (EBI-EMBL) (located in Hinxton, UK), the French National Center for Scientific Research (CNRS), the Foundation for Research and Technology in Hellas (Greece) and the University of Alberta in Canada.

Metabolomics is the study of the metabolites of a cell, tissue or organism. It contributes to the identification of small molecules with a molecular weight of less than 1,500 daltons, helping to better understand the biological system alongside omics technologies (genomics, transcriptomics and proteomics). Metabolomics studies rely on two strategies:

1) targeted analysis, which explores specific metabolic processes and the biomarkers thought to play a role therein;

2) non-targeted analysis, which aims to characterize all the metabolites contained in a biological sample, reflecting the overall composition of the metabolome.

The main analytical tools in metabolomics are mass spectrometry and nuclear magnetic resonance. Like other omics technologies, metabolomics deals with huge datasets. Univariate or multivariate statistical methods are used to identify chemical groups from molecular profiles. In practice, metabolomics is used for rognosis, diagnosis, patient stratification, monitoring of pharmaceutical treatment or surgery and biomarker discovery.

EXCEMET combines both analytical and computational expertise with the goal of strengthening the connections between metabolomics and biobanking. Carotenuto et al. show how the organization of the da Vinci European BioBank infrastructure integrates metabolomic information into pharmacogenomic research and personalized medicine [CAR 15].

2.3.4. *The FAIR principles of health data management*

Every day, physicians collect patient information and manually enter it into their electronic medical records. Compiling data from multiple centers and increasing the size of databases automatically increases knowledge and innovation in healthcare. Unfortunately, differences in clinical protocols and systems make it difficult to connect multiple sites. Before medical research problems can be solved, the first step is to ensure data integration. The study by Wilkinson et al. describes the principles of intelligent data management [WIL 16]. According to the authors, "good data management is not a goal in itself, but rather is the key conduit leading to knowledge discovery and innovation".

According to this study, there is an urgent need to reuse academic data. Scientific actors from academia, industry, funding agencies and academic publishers have defined the findable, accessible, interoperable, reusable

(FAIR) principles, which facilitate the reuse of data while respecting the privacy of the patient and the healthcare professional and medical confidentiality. The set of fundamental principles imposes that data should be findable, accessible, interoperable and reusable.

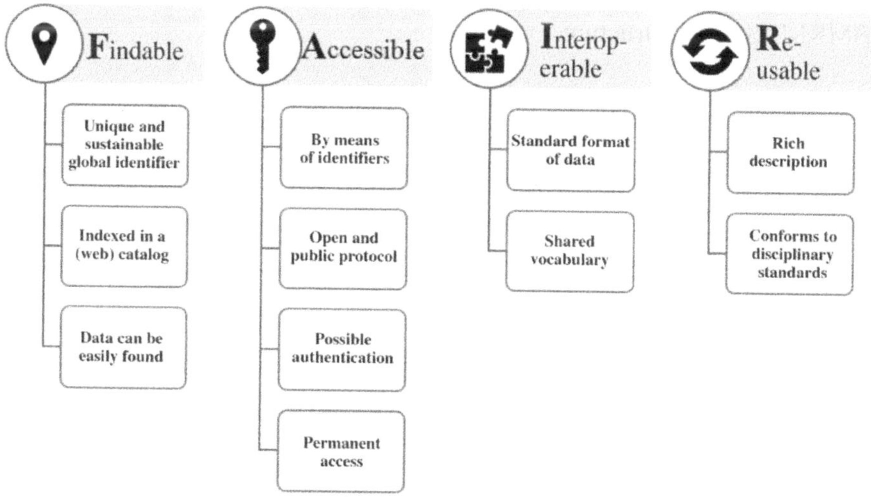

Figure 2.14. *FAIR principles for integrating and reusing the large amounts of information generated by scientific and medical research [WIL 16]*

High-quality samples are valuable to biobanks; however, more than the number of samples, it is the combination of data and clinical information linked to the samples that makes a collection strong. Medical research in the era of precision medicine is based on the analysis of samples linked to clinical data and, because correlations between results are often low, samples and data are tested on a very large scale. The more precisely the samples are characterized, the faster the research. This means that the needs of biobanks are constantly growing and the capacity for computerized data storage is also increasing. Modern biobanks are shifting from a sample-centric to a data-centric strategy [KOZ 19].

Researchers and physicians are always obtaining more data by increasing the size of their collections and multiplying their sources. From these large and heterogeneous datasets, new software is developed for data storage and analysis. For example, technological advances have decreased the cost of genomic analyses. Now, identification of genetic variants is affordable even on a large scale. This genomic testing capability has increased the need for data storage, and so have other omics technologies, as well as other kinds of sample biodiversity information, such as microbiome data. Several solutions have emerged to meet this need, such as cloud storage or "virtual" biobanks that bring together public organizations and private partners, like the BBMRI-ERIC infrastructure.

3

The Biobanking Lexicon

With the domestic and international proliferation of biobanks and their associated data, a common language for biobanks is essential.

[FRA 15]

The language of biobanks includes scientific, technical, regulatory and more specific terms. This chapter defines about 60 words or expressions useful for the biobankers' community, at the crossroads of several disciplines such as medicine, biology, computer science, law and economics. Its objective is to define the reference terminology essential to students of medicine or life sciences, as well as the users of biobanks.

3.1. Accreditation

From biospecimens, the biobanks proceed to the extraction of derived products. The high-quality samples, clinically annotated, are made available to the scientific community. The professionalization of biobanks aims at certification and accreditation, the two links in the chain of conformity assessment. However, what is the difference between the two?

Accreditation is a certificate issued by a *conformity assessment* body whose mission is to control inspection bodies. It is a formal recognition of technical and organizational skills.

Certification is a certificate relating to products, processes or systems issued by *control bodies* called certification bodies. It certifies the quality requirements for recommended procedures. In simple terms, certification requires mastery of the QMS, whereas accreditation requires not only mastery of the QMS, but also of technical skills.

The certification and accreditation standards for biobanks are national and international programs such as ISO 9001 (quality management systems), ISO 17025 (general requirements for the competence of testing and calibration laboratories) and ISO 15189 (requirements for quality and competence – medical analysis laboratories). Only *ISO standard 20387* is specific to biobanks (general requirements for biobanking), in line with the *French standard NFS 96-900* (quality of biological resource centers (BRCs) – management system of a BRC and quality of biological resources of human and microbial origin).

Biobank performance (proficiency testing (PT)) is assessed against pre-defined criteria through inter-laboratory comparisons. The purpose of PT is to measure the performance of specific measurements or tests and to encourage standardization of procedures to achieve consistent results. The evaluation is concerned with the sample handling procedures and the accuracy of the tests used in research. In short, the PT program evaluates the reproducibility and robustness of the methods used. A study by Gaignaux et al. showed that participation in *performance testing* is part of quality control for biobanks [GAI 16].

For *quantification tests* (assays), participants receive two or three samples and measure the requested parameters with a routine technique:

– DNA quantification (concentration in µg/mL) and purity (absorbance ratio at 260 nm/280 nm);

– RNA integrity using an Agilent, Biorad or other bioanalyzer;

– cell viability (%) using trypan blue or flow cytometry;

– tissue histology, quantifying the proportion of normal tissue and tumor tissue (notably the percentage of tumor cells) through histopathological examination of hematoxylin and eosin-stained slides from FFPE tumor sections.

For the *process tests*, participants receive a sample (a tube of whole blood and FFPE sections) on which the extraction steps are performed. The extraction products are sent to an external laboratory in charge of measuring the final concentration of DNA (and RNA) using spectrophotometry (in µg/mL), as well as its purity (ratio 260/280). The treatments concern the extraction of:

– DNA (and RNA) from whole blood;

– DNA from FFPE sections. The concentration of extracted DNA is expressed as µg of DNA per 20 µm of FFPE section).

By participating in PT programs, biobanks improve their performance, identify measurement and processing problems and build confidence among laboratory staff and users. Common errors are identified and corrected, which may be due to instrument calibration, operator training and sample dilution factors. For these reasons, PT programs are an important part of a laboratory's quality management system and are encouraged by ISO standards.

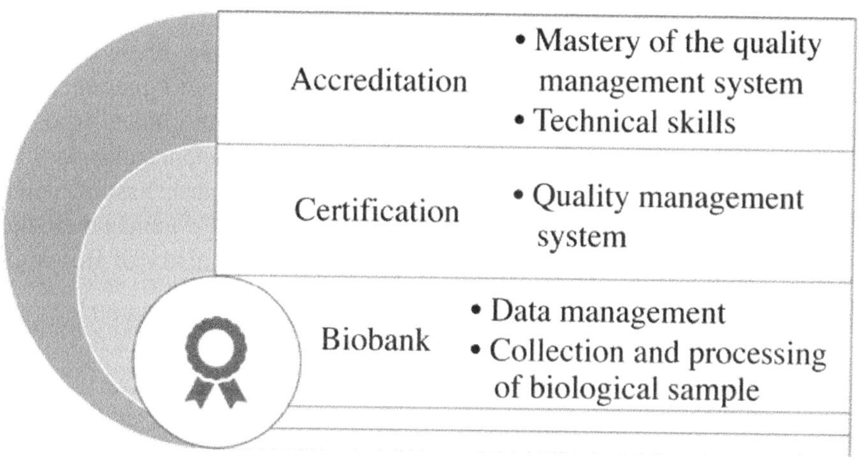

Figure 3.1. *Conformity assessment*

COMMENTARY ON FIGURE 3.1.– *Accreditation validates the performance of a biobank or any other organization, such as a certification body. Certification proves that the operating procedures follow predefined protocols for the quantification of biological data (measurements) and the processing of the sample (processes).*

3.2. Anonymization (or de-identification)

Biological sample management centralizes a vast amount of clinical, family, environmental and personal information that must remain confidential. The person's identity information – last name, first name and social security number – is encoded by a key or encryption code, assigned to all samples from the same individual. While absolute anonymization does not allow the return of clinical information to the patient, pseudonymization allows the re-identification of the patient. This coding facilitates the longitudinal follow-up of patients, which can extend over several years during an epidemiological study.

3.3. ANSM, the French National Agency for the Safety of Medicines and Health Products

The National Agency for the Safety of Medicines and Health Products (*Agence Nationale pour la Sécurité du Médicament et des produits de la santé* (ANSM)) is a French public institution. Its mission is to guarantee the safety of health products throughout their life cycles, from clinical trials to post-marketing monitoring (pharmacovigilance). Its competence also applies to biological products used for therapeutic purposes (organs, tissues, cells, blood products). Before starting a collection, the scientific manager or the biobank manager must apply to the ANSM and to the Ministry of Research for the conservation of the collection.

3.4. Artificial intelligence

Artificial intelligence enables quantitative evaluation of radiographic features. The development of algorithms through machine learning has enabled the automatic processing of medical images, improving cancer screening with a higher reliability than that of radiologists.

3.5. BBMRI, the coordinating center for European biobanks

Nearly 280 biobanks have joined the Biobanking and Biomolecular Resources Research Infrastructure (BBMRI) network. Its objective is to create a platform that connects researchers and allows scientists to share samples and their associated data. Initially, the goal of this infrastructure was to list the biobanks in Europe and to methodically identify all the collections. Then, the infrastructure was oriented towards the development of standard operating procedures, which are essential for harmonized routines when centralizing data in this unique network.

3.6. BBMRI-ERIC, the research infrastructure for biobanking

The objective of the Biobanking and Biomolecular Resources Research Infrastructure–European Resource Infrastructure Consortium (BBMRI-ERIC) is to bring together all biobanking stakeholders – researchers, biobankers, manufacturers and patients – to accelerate medical research and the discovery of new treatments. It is the largest European infrastructure dedicated to human health research. It helps researchers find the sample and data they need for their research project, while respecting the quality and regulatory constraints imposed on biobanks.

3.7. BBMRI-ERIC Biobank of Excellence or Expert Center

The main characteristic of a high-quality Expert Center is broad medical and scientific expertise. These centers have access to state-of-the-art technologies, a high level of standardization and professional quality management. Expert Centers interact with academic research centers and private companies for medical and pharmaceutical research. The BBMRI-ERIC network of Expert Centers links biobanks of expertise in different fields in different countries.

3.8. Biobank information management system (BIMS)

The biobank workflow provides patient information and sample data. From the patient's electronic medical record to the temperature control of the freezers, the biobank stores all the data using BIMS software, which ensures the traceability of samples and centralizes all the data associated with the sample. The software is hosted by a LAN that links several computers and analysis devices. The CDB and the HIS are connected to the BIMS, as well as the QMS and LIMS.

3.9. Biobank or biological resource center (BRC)

This structure ensures the management of biological samples – storage, preservation, processing and distribution – as well as the management of associated data. Biobanks can manage human, animal, plant or microbial samples. They are essential structures for scientific research. Collections of human samples are kept by structures historically called biological resource centers (a term consecrated by the OECD).

3.10. Biomarker

A *biomarker* is a quantitative indicator of a biological process or a therapeutic response. Molecular, histological, radiographic or physiological parameters are all biomarkers used in the search for therapeutic molecules. Others are used in routine clinical practice and some are used to determine the prescription of a specific treatment. A wide range of biomarkers are used today, alone or in combination.

3.11. Biospecimen

According to the American Cancer Institute, a biospecimen is defined as "A sample of material, such as urine, blood, tissue, cells, DNA, RNA, or protein, from humans, animals, or plants. Biospecimens may be used for a laboratory test or stored in a biorepository to be used for research" [NAT 11]. Regarding animal and plant samples, the purpose of the collections is to preserve the biodiversity of the species, as illustrated in the Frozen Ark project and the Svalbard seed reserve.

The Frozen Ark project, named after Noah's Ark, aims to preserve the DNA and cells of endangered species [THE 20a]. Initially dedicated to the preservation of a small snail in danger of extinction, the project has evolved towards the preservation of all animal species in danger of extinction. Created in 2004, this biobank is headquartered at the University of Nottingham, one of its co-founders along with the Natural History Museum and the Zoological Society of London. The Frozen Ark consortium partners are spread across five continents and include universities, zoos, natural history museums and research institutes. However, if global biodiversity is declining, what is the point of conserving the heritage of endangered species? The goal is not to recreate the organism from the preserved material, but rather to preserve a wealth of information about biology, ecology and evolutionary history that would be lost forever [BRE 18].

In order to preserve plant species, the Svalbard global seed vault was established in 2008 in the Norwegian archipelago of Svalbard in the Arctic Ocean. The dried and vacuum-packed seeds are stored at low temperatures (–18°C) for long periods. "The seeds are preserved more easily and remain viable for about 50 years, the aim being to regenerate entire plants", says Jean-Louis Pham, a researcher at the *Institut de recherche pour le développement* (IRD) and leader of Agropolis Resource Center for Crop Conservation, Adaptation and Diversity (ARCAD). This project is devoted to the development of a multi-institutional center for conservation of and research on Mediterranean and tropical plants in Montpellier.

The Svalbard Seed Reserve is funded by the Norwegian government and managed by the Norwegian Ministry of Agriculture and Food, the Global Crop Diversity Trust and the Nordic Genetic Resource Centre (NordGen).

After 10 years of operation, the number of samples has reached 1,060,987 seeds, deposited mostly by international agricultural research centers [ASD 18]. Among the 5,000 different species, rice and wheat represent more than 150,000 species each. These seeds, already stored in university gene banks or private companies, are stored in this "plant Noah's Ark" for security.

For human health, there are a wide range of samples, the main ones being: whole blood, saliva, tissue, red blood cells, buffy coats, plasma, serum and urine. They are grouped into several types of collections:

– *DNA banks* in which DNA has been extracted from blood or cells. They are used to search for mutations and immunogenetic and pharmacogenetic predisposition factors;

– *biological fluids* include plasma, serum, urine and cerebrospinal fluid. They are used for serological and immunological markers;

– *primary cultures* include cancer cells. These lines are studied in molecular genetics;

– *fixed tissues* are used for anatomopathological, genetic and cell biological studies.

In Lee et al.'s study conducted from 1977 to 2016 on the types of samples mentioned in scientific publications, it was found that the most used samples are tissue samples [LEE 18]. The study shows that *microbiome* publications have been rapidly increasing since 2010 and are overtaking less represented samples, such as *peripheral blood mononuclear cells* (*PBMCs*) and buffy coats.

Among the publications containing serum, plasma and tissue samples, which represent the majority of samples, the analyses performed are mostly related to proteins and RNAs, to the detriment of DNA, free genomic DNA or metabolites.

Common techniques are mass spectrometry and next-generation sequencing (NGS).

In the biobanking lexicon, other terms can be used to refer to the biospecimen depending on the stage of processing:

– the *biological sample* corresponds to a portion of the biospecimen (initial sample). This fraction is analyzed or stored;

– the *derivative or derived product*, such as DNA or RNA, is obtained after sample processing, for example nucleic acid extraction;

– an *aliquot* is a fraction of a sample or derivative.

3.12. Buffy coat

After centrifugation of whole blood, the buffy coat is located at the interface between the high-density red blood cells (lower phase) and the lower density plasma (upper phase). This fraction enriched in leukocytes and platelets is used in blood transfusion to produce platelet pools. Biobanks store these samples, which are part of whole-blood products such as plasma, serum and nucleic acids (DNA).

3.13. Certification

Certification refers to a certificate relating to products, processes or systems, issued by an inspection body or certification body. It certifies that the quality requirements have been carried out in accordance with recommended procedures. Certification standards are national programs – such as the *French national standard NFS 96-900* describing the quality of BRCs, the management system and the quality of biological resources of human and microbial origin. As an international extension, the *ISO 20387 certification* specific to biobanks sets out the general requirements for biobanking.

3.14. Clinical biobank

The specimens and their associated information come from patients with a specific disease, hence the term "thematic biobank". Collected during the patient's medical diagnosis or treatment, these pathological samples may be biological fluids (blood and urine) or tissues. They are often stored at the place of collection in a hospital biobank. Clinical biobanks are dedicated to a given pathology, aiming to understand its molecular and cellular mechanisms and to discover therapeutic avenues.

3.15. Clinical research or clinical trials

Clinical research is a key stage in drug development, carried out in partnership with hospitals. It is organized in several phases during clinical evaluations (or trials) carried out on human volunteers in order to investigate the safety of the drug on a small number of people (phase I), the therapeutic efficacy of the product on a large number of patients (phase II), the benefit/risk ratio (phase III) and finally, after marketing, the large-scale follow-up (phase IV).

3.16. Clinical trial sponsor

In order to protect participants in biomedical research, French law no. 2004-806 on public health, known as the Huriet–Sérusclat law, imposes responsibility on the *sponsors* of clinical trials. The concept of a sponsor identifies the individual or legal entity who takes the initiative in medical research. They are responsible for setting up, managing and financing the study. The sponsor entrusts the study to an experienced "investigator" physician, an individual who supervises the project. For multicenter trials, several investigators are recruited.

3.17. CNIL, the French National Commission for Information Technology and Civil Liberties

In French law, the protection of personal data is legally regulated by the law relating to data processing, files and freedoms (law of January 6, 1978, as amended), which the French National Commission for Information Technology and Civil Liberties (*Commission nationale de l'informatique et des libertés* (CNIL)) is responsible for enforcing. The CNIL must be consulted before any data processing is carried out in order to ensure security and confidentiality, as well as the proper use of personal data, especially if it is genetic data.

3.18. Cohort and megacohort

Multifactorial studies are long-term and require the inclusion of hundreds of thousands of participants to detect the onset of a disease. Epidemiological cohorts are designed to study and monitor populations over long periods of time (10–15 years) in order to monitor the development of various diseases. The goal is to identify genetic or environmental factors that lead to the development or progression of a disease.

3.19. Collection of biological samples

BRCs are "[…] service providers and repositories of the living cells, genomes of organisms, and information relating to heredity and the functions of biological systems […] BRCs must meet the high standards of quality and expertise demanded by the international community of scientists and industry for the delivery of biological information and materials", as defined by the OECD.

The constitution of a collection of human biological samples for genetic or non-genetic research purposes is regulated by articles L. 1131-4, L. 1243-3 and L. 1243-4 of the Public Health Code. Once the collection project has been declared to the competent administrative authority, the latter ensures that the conditions of constitution, conservation and exploitation of the collection present sufficient guarantees to ensure the proper use, security and confidentiality of the data collected.

3.20. Companion test

The identification of new *biomarkers* has enabled the development of tests for the use of targeted therapies. These are diagnostic tests that make it possible to select only those patients in whom the treatment is likely to be of benefit. For this reason, such a test is referred to as a "companion" test.

3.21. Computed tomography (CT)

This imaging technique, also called a CT scan or scanner, is a common examination during cancer screening and then during patient follow-up. This technique is based on the principle of X-ray absorption in the volume of tissues and organs. CT is used to evaluate structures but does not provide information on their functionality or molecular composition. In contrast, functional imaging explores physiological processes using dynamic contrast-enhanced magnetic resonance imaging (DCE-MRI), providing in vivo vascular information, for example.

3.22. CPP, the French Committee for the Protection of Individuals

In order to protect participants in biomedical research, law no. 2004-806 on public health, known as the Huriet–Sérusclat law, requires that clinical research protocols be submitted to an advisory committee for the protection of individuals. This independent body, which has a legal character, is composed of representatives of patient or health system user associations (approved under article L. 1114-1 of the French Public Health Code). Its members are appointed by the legal representative of the State in the region where the CPP is located. The favorable opinion of the CPP is mandatory to start a research project.

3.23. Data protection officer (DPO)

At the European level, the new GDPR requires the designation of a DPO to manage data protection. He/she has the duty of ensuring fair and transparent data processing, notably in biobanks. A biobank's DPO can be contacted with any questions relating to the biobank's approach to data protection and the GDPR. He/she registers the data processing operations involving personal data (data collection, data generation, data sharing, informed consent). With the increasing use of new technologies, he/she takes into account the privacy rights of individuals. The DPO is the safeguard of data security and confidentiality as mandated by the Data Protection Act (DPA) in common law.

3.24. Diagnostic biomarker

This type of biomarker is used to identify carriers of disease. Disease classifications are now moving towards a molecular basis in association with medical imaging. Any radiology test that requires measurement and evaluation can be considered a diagnostic biomarker.

3.25. Digital health or e-health

Over the past 20 years, digital transformation has made its entrance into biobanks and more broadly into healthcare with the rise of e-health. Artificial intelligence helps in medical imaging diagnosis with learning programs that are more reliable than the analysis of images by radiologists (lung cancer). It is capable of early diagnosis (preventive medicine), screening billions of molecules and identifying the biomarkers of a disease, and of discovering new drug candidates, setting up new connected medical devices thanks to clinicobiological data.

3.26 Electronic case report form (eCRF)

In clinical research, the data collection is recorded in an electronic observation book. It includes all the individual information collected during the study: patient identification, inclusion criteria, informed consent and all the results of medical examinations in compliance with regulatory and ethical constraints. The exchange of information by research teams is now done electronically, facilitated by the implementation of the EHR in most American hospitals or the *dossier médical partagé* in France.

3.27. ESBB, the European, Middle Eastern and African Society for Biopreservation and Biobanking

The mission of this network of biobanks is to advance the sharing of biological resources in health, education and the environment in the countries of the Europe, Middle East and Africa (EMEA) region.

3.28. FDA, the US Food and Drug Administration

This department of the US government controls food and pharmaceutical products. It exercises regulatory control over drugs, diagnostic kits and medical devices before they are marketed.

3.29. Formalin-fixed, paraffin-embedded (FFPE) tissue

FFPE blocks play an important role in biobanks because they are easy to store compared to fresh tissue. FFPE blocks are dehydrated tissue fixed with formalin and embedded in paraffin. They are stored at room temperature as solid blocks that are easy to handle and cut. With FFPE sections, hematoxylin and eosin staining (HES), analyzed by pathologists, is used to validate the histopathology of the tissue and the presence of cellular damage.

Formalin fixation and paraffin embedding preserve the morphology and cellular structure of tissues. Thus, this technique has become the standard method of preservation for medical and surgical diagnosis. Long-term storage of FFPE blocks at room temperature is less expensive than freezing at very low temperatures. Pathology laboratories store large numbers of FFPE blocks compared to frozen tissue. These often-untapped resources represent quantities of material that can be used for long-term clinical follow-up.

FFPE blocks are used in immunohistochemistry for disease diagnosis and in molecular biology for nucleic acid and protein extraction. The work of Kokkat et al. showed that the quantity and quality of DNA extracted from three groups of blocks stored for different lengths of time – 11–12 years, 5–7 years and 1–2 years – were comparable to those of DNA extracted from blocks less than one year old [KOK 13]. Similar results were obtained with proteins extracted from all three groups. In this study, frozen tissue was used as control tissue. Nucleic acids and proteins were extracted from FFPE blocks. Even though protein fragmentation was higher in FFPE tissues stored for more than five years, the positive control (GAPDH) was present in all tissues.

These results show that despite variability in collections and treatments, FFPE tissue blocks stored for more than 12 years can be used for DNA and protein extraction. Formalin fixation and freezing preserve DNA, but only freezing preserves RNA.

3.30. Free and informed consent

By signing an informed consent form (ICF), an individual expresses their willingness to participate in a research project (inclusion). This consent, which is sent to the CPP, is mandatory when it comes to the use of biological samples and personal data. Electronic informed consent (e-consent), which is already regulated by the guidelines of the FDA but non-existent in France, refers to the use of electronic systems, which could be in the form of audio or video media, to transmit information about a study in order to obtain and document informed consent.

3.31. General Data Protection Regulation (GDPR)

At the European level, the new General Data Protection Regulation reinforces the principles of data protection and the rights of citizens. This regulation applies to activities of the biobank both upstream, for the tasks related to the collection of personal data, and downstream, for the creation of new data from and associated with the samples. The processing of personal data must comply with French law no. 78-17 of January 6, 1978, relating to "information technology, files and freedoms", consolidated version from July 23, 2019. The CNIL must be consulted before any data processing to ensure that security and confidentiality are respected, as well as the proper use of personal data, especially in the case of genetic data.

3.32. Imaging biobank

The *medical imaging biobank* is a "structured repository of imaging data and biomarkers from radiology and beyond, shared among researchers and linked to other biomedical banks", according to the European Society of Radiology. Traditional and imaging biobanks work together in the context of personalized medicine.

3.33. Imaging biomarker

Medical imaging techniques make it possible to explore the entire body or to visualize a particular organ during a routine examination. By measuring a volume or a diameter, this biomarker allows the detection of early signs, even before the appearance of clinical signs. These biomarkers are applied in various investigations, such as into the distribution of adipose tissue, cerebral and cardiac functions and abdominal and thoracic organs.

3.34. ISBER, the International Society for Biological and Environmental Repositories

The mission of this network of biobanks is to preserve and conserve collections of plants, seeds, animal organisms and microorganisms from the environment. The ISBER Good Practice Guide shares strategies recognized by the scientific community and assists biobank users in the accreditation process according to national or international standards.

3.35. ISO 20387:2018 Biotechnology – biobanking – general requirements for biobanking

This international ISO standard is specific to biobanking activities and describes the requirements for competence, impartiality and consistent operation in biobanks. ISO 20387:2018 applies to all organizations that perform biobanking for research and development. It recognizes biobanking activity among biobank users and regulatory authorities. Biobanks that receive this label give users confidence in the overall functioning of the biobank and therefore in the samples transmitted. In addition, this label increases the international visibility of the biobank.

3.36. Laboratory information management system (LIMS)

The processing of the sample in the biobank is supervised by the LIMS software at each step: from reception to preparation and storage and finally distribution of the sample or its associated data. This software facilitates the standardization of procedures and ensures the control of the process.

3.37. Liquid biopsy (in oncology)

A biopsy is a method of cancer diagnosis. It involves taking a sample of tumor tissue and adjacent healthy tissue and examining them histologically for the presence of cancer cells. As an example, *tissue biopsy* is a delicate invasive technique for tumors that are small or located in the lungs.

An alternative to tissue biopsies has been proposed for the detection of diseases. This technique consists of taking a biological fluid and looking for biomarkers of disease. This new application is called *liquid biopsy*, and it focuses on samples of blood, saliva, urine and pleural and cerebrospinal fluid. Unlike tissue biopsies, it is not subject to interpretation limitations due to intratumoral heterogeneity. It is non-invasive and facilitates the detection of *circulating biomarkers*. They can be detected before the first symptoms of disease. This is the case with *circulating tumor cells* (CTCs), which can be detected as soon as they enter the bloodstream, circulating tumor DNA fragments (ctDNA) released by apoptotic cancer cells, microRNAs and exosomes, vesicles released by CTCs and protein biomarkers.

In order to identify CTCs as early as possible, Paul Hofman's team developed a blood test based on the isolation by size of tumor cells (ISET) procedure in collaboration with the company Rarecells Diagnostics [ILI 14]. This technique separates CTCs according to their size. By filtration of the blood, CTCs with a size of 15–18 microns, sometimes up to 30 microns, are separated from smaller blood cells (between 7 and 8 microns). The study involved a group of 245 volunteers, 168 of whom had chronic obstructive pulmonary disease (COPD), a known risk factor for lung cancer. The participants underwent blood tests and various imaging tests. Via the blood test, CTCs were identified in five people (3%), while imaging revealed no lung nodules. These patients were followed for four years: five of them developed a nodule, detected by CT scan. They were immediately operated on and histopathological analysis of the nodule confirmed the diagnosis of early stage lung cancer. Prior to the imaging detection of symptoms, so-called "sentinel" CTCs detected cancer in the five patients who triggered the disease. The validation stage for this blood biomarker requires confirmation of the result in a more representative sample group.

As part of the Air project on early detection of lung cancer, Professor Paul Hofman's team followed a cohort of 638 at-risk smokers or ex-smokers with COPD. They were examined with a thoracic scanner and blood tests every year for three years. The results were published in 2020 in the scientific journal *Lancet Respiratory Medicine*, and the professor describes the interest of combining blood tests and CT scans [MAR 20]. "On an image, we see spots but we don't know if the nodules are cancerous or not. The whole point of the blood test is to allow the thoracic surgeon to be sure that what he is going to operate on is really cancer, and not another lesion, which is extremely frequent. It is important to know that the thoracic scanner gives 30% false positives", reports Paul Hofman. In fact, the ISET method had an overly low sensitivity for detection of patients at high risk of developing lung cancer, even though the specificity was high [MAR 20].

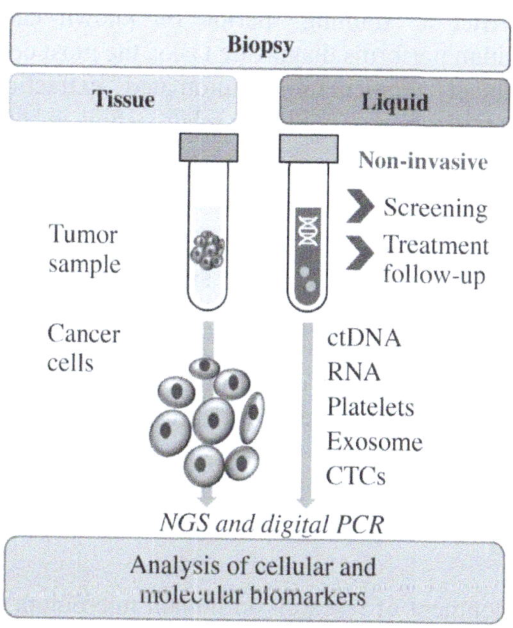

Figure 3.2. *Applications of tissue and liquid biopsies in oncology*

COMMENTARY ON FIGURE 3.2.– *Cancer cells that break away from tumors are found in the bloodstream, along with fragments of cancerous genetic material. ctDNA, CTCs and the exomes they secrete act as biomarkers that can be detected by NGS, RT-PCR and digital PCR, in the case of ctDNA. Liquid biopsies are a new screening tool (early detection of cancer), essential for patient stratification (therapeutic intervention) and real-time treatment monitoring in the face of the risk of metastatic relapse.*

3.38. Machine learning and deep learning

Machine learning utilizes programming methods based on experimental data. These algorithms are able to learn from the data and improve autonomously. Deep learning, a subcategory of machine learning, is the most developed form of artificial intelligence in medical imaging. The technology is based on CNNs. It can use supervised or unsupervised human learning methods. In practice, two sets of patient metadata are processed according to two different schemes. With the first dataset, deep learning models are developed and evaluated. Network validation is then performed on the second dataset. After a "training" period on known data (the training dataset), the algorithm performs diagnostic tasks, the most common of which are detection, object segmentation, automated extraction of imaging descriptors and identification of nonlinear relationships in complex data.

3.39. Medical imaging

Medical imaging has become more sophisticated with the entry of artificial intelligence, which has been used in several clinical image acquisition techniques such as CT, PET and MRI. Information from biological samples crossed with data from medical imaging has allowed medicine to take a quantum leap forward in the form of statistical algorithms.

3.40. Microbiota

With the development of research on human microbiota, more and more biobanks have built up collections of microbiota specimens.

These human microbiota combined with genetic information represent a source of valuable health data. However, what are microbiota or the *microbiome*? Both terms are often used.

According to the vocabulary of microbiome research compiled by Marchesi et al., "The microbiome is the entire community of microorganisms (i.e. bacteria, archaea, fungi, protozoa, phages and viruses) with their genomes living in a particular environment (e.g. in/on a human/animal body/body site, plant, soil)" [MAR 15]. The difference between the terms "microbiota" and "microbiome" lies in the specialization towards the genomes of microorganisms entailed by the term microbiome.

As with tissue samples, the types of specimens collected and the associated clinical data vary depending on the purpose of each biobank.

For example, the American Human Microbiome Project has collected samples via the American Gut data platform to characterize the microbial diversity of the American population and understand how diet and lifestyle can contribute to health through the microbes in the gut of each individual [THE 20b]. In Europe, the Meta-genomics of the Human Intestinal Tract (MetaHIT) consortium has established that the microbiome comprises 3.3 million genes (150 times more than the human genome). This work paved the way for the exploration of the microbiota by allowing the rapid identification of the microbes present in its composition from their genes.

Stool banks are poised to grow in response to the treatment of intestinal diseases, particularly *Clostridium difficile* infection. According to the FDA, a stool bank is defined as "An establishment that collects, prepares and stores FMT products for distribution to other medical institutions, healthcare providers, or other entities for therapeutic or clinical research."

In the United States, the first OpenBiome stool bank was established in 2012 in Massachusetts. Its objective is to eliminate the practical obstacles in *fecal microbiota transplantation (FMT)* by providing safety and standardization for the process and, in parallel, engage in translational research into the human microbiome.

In France, the Lyon-based biotech Maat Pharma – which stands for "Microbiota as a Therapy" – was founded in 2014. It specializes in the transplantation of human gut microbiota (microbiotherapy) to rebuild the microbiome after cancer treatments or after significant administration of antibiotics.

To provide samples for microbiota research, biobanks need to standardize pre-analytical variables that can influence composition and analysis. The main parameters are donor age, gender, diet, stress, smoking, medical follow-up and lifestyle. In addition, the microbiota depend on the location of the sample in the body and the donor's health status.

As soon as the sample is taken from the organism in view of being stored in a collection, the composition of the microbiota may change since the habitat of the microorganisms is modified (temperature and oxygen). The sample must be immediately stabilized or frozen. Contamination of non-microbial DNA, and therefore of the host, is an additional parameter to be controlled during microbiota sequencing.

The importance of the pre-analytical phase has led to the standardization of the collection and preservation of microbiota samples through an international project called the *International Human Microbiome Standards* (IHMS) [SAN 14].

Subsidized by the European Commission, the IHMS project provides access to standard procedures and promotes exchanges between sample users.

This program is articulated around four axes:

– international coordination and project management;

– standardized sample collection and processing;

– sequencing standards;

– data analysis.

Transplantation of fecal microbiota opens a therapeutic avenue for the treatment of inflammatory bowel diseases (IBDs) and irritable bowel syndrome (IBS).

A personalized approach is required for each of the process steps, as described by Benech et al. [BEN 20].

First, candidates for FMT must be selected using clinical, microbiological, immunological and metabolic criteria based on blood and fecal analyses (microbiota diversity). Second, the FMT procedure must be established for the type of transfer (oral or injection) and the characteristics of the donor. Finally, the last step consists of monitoring the effectiveness of the FMT through the measurement of clinical, biological and immune parameters and the composition of the microbiota.

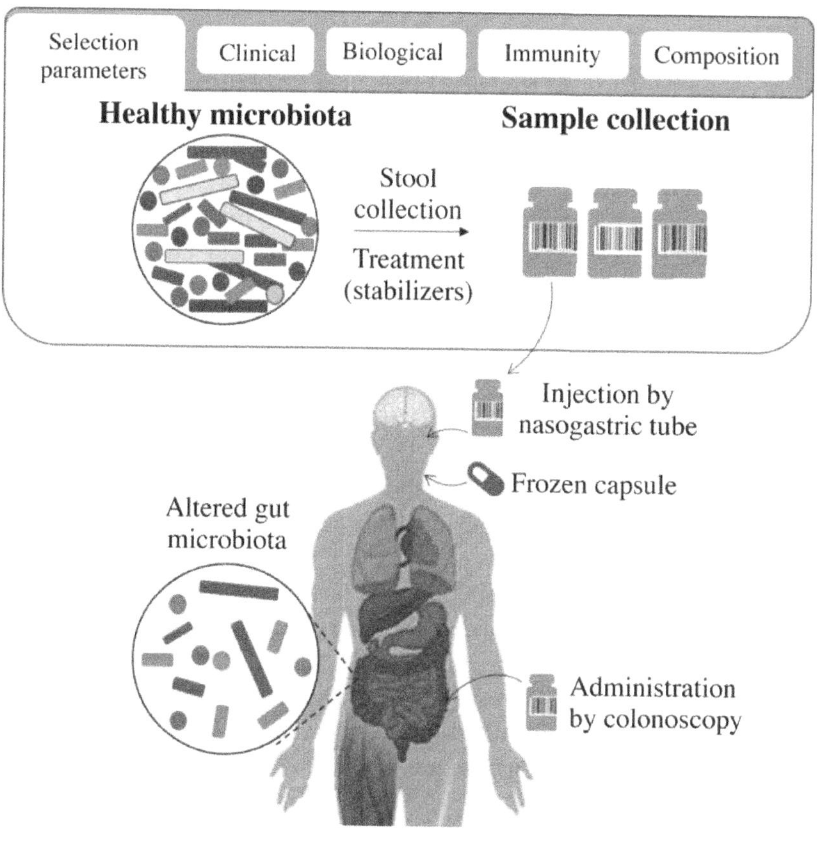

Figure 3.4. *The process of fecal microbiota transplantation*

COMMENTARY ON FIGURE 3.4.– *Donor selection is based on clinical and medical workup, blood work and stool analysis to confirm microbiota composition and absence of pathogens. Administration can be undertaken by colonoscopy, through a nasogastric tube or by oral administration of frozen capsules.*

3.41. Monitoring biomarker

Routinely measured, these biomarkers assess the status of a disease or exposure to a medical or environmental product. Exposure biomarkers are very important to assess risk and ensure the safety of participants during clinical trials by making it possible to routinely monitor liver toxicity and the cardiotoxicity of drugs. They are also used to measure pharmacokinetic effects and to detect early therapeutic responses and disease complications.

3.42. MTA or biological material transfer contract

The need for sample collections is increasing with the multiplication of medical research projects. This is especially true for rare diseases. However, this growing research potential does not correspond to an increase in sample sharing, which is hampered by international legal and ethical regulations. The study by Mascalzoni et al. develops an international data and sample transfer tool that meets the specific demands of standardization and transparency in biobanks [MAS 16].

MTAs or data transfer agreements (DTAs) are contracts that determine the use of samples and data, the link to the consent and the duration of the contract. It is fundamental to ensure that the sharing of samples and data is undertaken within an ethical and legal framework for donors and research partners. Biobanks, upon request, can make samples and data available to researchers for any current or future research project. The transfer of samples and data to other biobanks requires the approval of the biobank management, the prior authorization of the ethics commission, and the signing of the *material transfer agreement*. The transferred samples and data are coded before being transmitted to the researchers. The key to the code is kept by the issuing biobank in a secure manner.

The MTA is an agreement whereby a biobank transfers biological material to a partner (academic or industrial). The MTA governs the use of this material while defining the rights and duties of each party.

The MTA must describe the types of data and samples released to the researchers and their use in the research project. The biobank must be committed to ensuring the quality of the data and samples in accordance with international standards. If the transferred material is accompanied by complex data exchange, reports and scientific discussions, then the contract must be reclassified as a research collaboration contract. In concrete terms, the MTA amounts to establishing a material transfer agreement and monitoring the results of the study conducted on the material.

Faced with the need to regulate exchanges between biobanks, Cervo et al. proposed a ready-to-sign MTA model [CER 16]. Frequently, researchers do not use the MTA because they are not familiar with the procedure. Cervo et al. created a standard template and identified the main information that should be included in the contract between two biobanks. In particular, transfer contracts should mention:

– the identity of the persons responsible for the sending and receiving biobanks;

– the purpose of the contract by describing the collection and the research project;

– safety instructions for the use of the materials;

– data confidentiality;

– the intellectual property protection for sending and receiving biobanks.

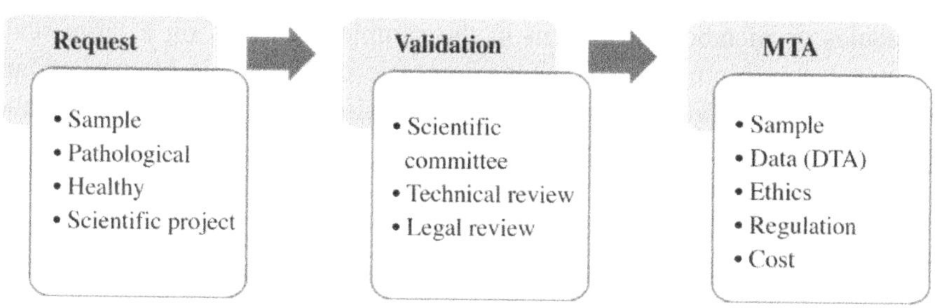

Figure 3.5. *The procedure for a material transfer agreement or data transfer agreement*

COMMENTARY ON FIGURE 3.5.– *The request specifies the nature and number of samples requested and whether they are pathological or healthy and essential to the scientific is project briefly described. This request is validated by a scientific committee from the issuing biobank, which draws up the contract in compliance with legal and ethical regulations.*

3.43. NF S96-900

In France, the NF S96-900 standard combines quality management with technical skills in the field of biobanks. It is specific to structures that undertake BRC activity. This standard provides requirements for the quality management systems of BRCs in order to facilitate the acquisition, validation, conservation and distribution of biological resources by guaranteeing the quality assurance and traceability of samples. To meet the criteria of these standards, these structures can rely on a QMS that meets the requirements of ISO 9001 certification, based on the implementation of standardized procedures and audits.

3.44. Omics

Through the use of high-throughput technologies, omics has generated massive data. These megadata come from various types of biological samples (tissue, blood, urine, cells, nucleic acids, etc.) and the associated medical data. Omics technologies allow researchers to develop genomic (DNA sequencing), transcriptomic (all transcribed messenger RNA), proteomic (analysis of proteins in the sample) or metabolomic (small molecules or metabolites present in the sample) analyses on a large scale, making it possible to identify new prognostic or diagnostic biomarkers and new therapeutic targets. In practice, the analysis of data from a large number of patients (Big Data) is evolving towards the analysis of a very large volume of data from a single patient (Fat Data).

3.45. Peripheral blood mononuclear cells (PBMCs)

Blood mononuclear cells have a single nucleus. They include the mononuclear leukocytes lymphocytes (T cells, B cells and NK cells) and monocytes. They can come from adult peripheral blood or fetal cord blood. They are isolated from blood samples by a density gradient (Ficoll). They are stored in biobanks along with whole blood, plasma and nucleic acids from patients with blood diseases (hemopathies).

3.46. Personalized or individualized medicine

Personalized medicine has transformed the one-size-fits-all healthcare offer into an individual approach from diagnosis to treatment. Omics technologies provide a complete view of the molecular and metabolic biology data which characterize the individual. Only biobanks provide access to large-scale samples offering a wide range of genetic, proteomic, transcriptomic and metabolomic analyses.

3.47. Pharmacodynamic/response biomarker

These biomarkers are biochemical, histological and anatomical parameters that assess the response to therapy. For many drugs, variation in the pharmacological biomarker results in a change in dose during therapy.

3.48. Population biobanking

Samples are collected from non-selected volunteers from the general population. Blood samples are collected along with associated participant information such as medical and family history, lifestyle and environmental exposure. The tests are then followed for several years, even several generations. Through population biobanking, it is possible to identify predictive biomarkers to understand the factors influencing the onset of complex diseases. This prospective approach is the strength of medical prevention programs.

3.49. Positron emission tomography (PET)

Also called positron emission tomoscintigraphy or a PET scan, this is the least used technique because its resolution is inferior to that of a CT scan. This examination is based on the injection of a weakly radioactive tracer into the blood. For example, radioactive glucose is metabolized more rapidly in the tumor cells and provides functional information on the metabolism and proliferation of the tumor.

3.50. Pre-analytical phase

Sample integrity requires control of the pre- and post-sampling process through to storage. Hence, it is important to control *pre-analytical variables* such as the nature of the sample and the storage technique and its duration. The nature of the sample must represent the sampling site (organ) and the physiopathological state of the patient (healthy or suffering from a pathology). The preservation technique must be adapted to the type of sample, solid or liquid. A tissue sample is fixed in formalin and embedded in a paraffin block at a controlled temperature (between 55°C and 58°C), whereas a fluid sample is frozen in the presence of a cryoprotectant (glycerol or dimethylsulfoxide). Harmonization of protocols is implemented at each step of the pre-analytical chain under the pathologist's supervision, starting with the surgical procedure, followed by the transport of the sample to the biobank and finally its conservation.

3.51. Precision medicine

Precision medicine provides a therapeutic avenue based on the genetic and environmental profile of the individual. Omics technologies characterize the individual first by:

– genomics: the sequencing of the individual's DNA (genome);

– transcriptomics: all the transcribed messenger RNAs;

– proteomics: the analysis of proteins in the sample;

– metabolomics: all the small molecules (metabolites) present in the sample.

In addition, two other factors complete the individual's profile:

– the microbiome: all the microorganisms present in the body;

– the exposome: all the individual's relationships with the environment.

This complex dataset is a haven for precision medicine studies to mimic the individual on the one hand and their membership in a population and one of its subgroups on the other hand.

3.52. Predictive biomarker

A predictive biomarker allows the targeting of specific groups of people for whom a particular drug may be useful. It correlates with the therapeutic activity of a drug. The diagnostic test associated with the treatment is called the "companion" test. To date, predictive biomarkers are primarily genetic variants. The main field of targeted therapies – stratified – is oncology.

3.53. Prognostic biomarker

A prognostic biomarker is used to identify the likelihood of a clinical event, such as disease recurrence or disease progression. It reflects the pathophysiological process of the disease, independent of drug treatment.

3.54. Prospective study/survey

An example of a prospective study is the Rotterdam Study, a prospective cohort study started in 1990 in the city of Rotterdam in the Netherlands. Its primary objective is to study the aging of the population by targeting cardiovascular, endocrine, liver, neurological and pulmonary diseases. Since 1990, participants in the study have been filling out a health questionnaire (medication use, medical history) and describing their lifestyle habits (smoking, alcohol consumption, physical activity). Every three or four years, they undergo a complete check-up including medical imaging and blood sampling for molecular and genetic analysis.

3.55. Quality management system (QMS)

The QMS defines the quality strategy of a biobank and the objectives for quality assurance (QA) and quality control (QC). This requires the control of the sample process from its reception to its transfer, allowing its traceability. Sample quality control indicators, such as sample integrity, ensure the reproducibility of results. In addition, clinicopathological annotations specify the characterization of the sample, which then becomes a "qualified" specimen. An efficient QMS is necessary to obtain accreditation for the biobank and thus to gain recognition from the scientific community.

3.56. QR code

The samples are listed in the form of a QR code (2D data matrix code) that contains the minimum information about the patient: identification number, age, gender and primary diagnosis. In addition, detailed information is entered into the hospital's clinical information system in connection with the sample identification.

3.57. Radiomics

Radiomics is the extraction of a large number of quantitative parameters called descriptors (more than 200) from a medical image. The hypothesis is that automatic quantitative analysis by software provides better information than visual recognition by the radiologist. All this information is considered as an expression of the patient's *phenotype* and can be correlated with the gene expression profile, hence the term *radiogenomics*.

3.58. Region of interest or volume of interest (ROI/VOI)

In medical imaging, regions or volumes of interest involve manually or automatically delineated areas such as tumor volume measurements, quantification of metabolic activity in the ROI and image characteristics (texture and shape).

3.59. Retrospective study/survey

The discovery of new candidate biomarkers leads to retrospective studies, enabled by large sample collections and accompanied by accurate epidemiological, clinical, histological and genomic data. These data can be shared in a national or international network of several biobanks.

3.60. Standard operating procedures (SOPs)

A high *level of standardization* is achieved by implementing harmonized operating procedures and certification and accreditation procedures common to the biobank network. Biobanking Expert Centers are a resource for the biotech and pharmaceutical industries. They implement state-of-the-art technologies, participate in ringtest evaluations and publish SOPs for pre-analytical steps, molecular analysis and complex data management.

3.61. Stratified medicine

Twenty-first-century medicine responds to the patient's profile, the type of disease and the environment in which they live. Research has "stratified" groups of patients according to their genetic profile in order to individualize therapy and deliver tailored treatment to patients who are likely to respond to treatment.

3.62. The process of fixation and paraffin embedding

Fixation is intended to immobilize cellular and tissue structures. It must not create artifacts and must ensure the preservation and faithful image of the tissue and cells.

The objective of *dehydration* is to replace the water in the tissues by paraffin. The paraffin being immiscible with water, the tissues have to be dehydrated in baths of alcohols and xylene and finally in paraffin heated to its melting point. When the hot impregnation is finished, the pieces are put in blocks where they can be kept indefinitely.

The dehydration and inclusion process is automated and involves three steps:

– the dehydration is undertaken progressively using baths of alcohols of increasing concentrations (70°, 80°, 95° and 100°), so that the fluids substitute each other until the impregnation in the paraffin;

– clarification must be undertaken with xylene because paraffin is not miscible in alcohol;

– paraffin impregnation is then undertaken, usually without DMSO, with a melting point between 56°C and 58°C.

The duration of the different steps varies according to the type of tissue and the sample size. Once the pieces are impregnated with paraffin, they must be included in a paraffin block to allow for *microtome* cutting. It is the embedding that is the critical step for the final orientation of the sample in the block and thus on the slide after the cut.

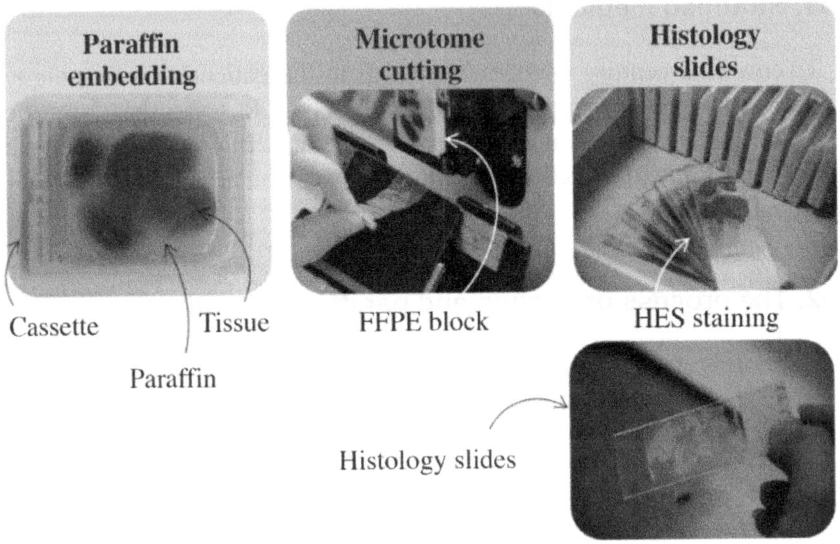

Figure 3.3. *Formalin-fixed, paraffin-embedded tissue blocks*

COMMENT ON FIGURE 3.3.– *Cut into thin sections with a microtome, the tissues are placed on histology slides and then stained with hematoxylin and eosin (HES) to visualize tissue and cell morphology and localize lesions. White, unstained slides are retained for future staining. The images here were taken from the website of the University of Poitiers' morphological research platform [CHU 20].*

3.63. Translational research

Translational research provides the link between academic research and clinical research, in other words between research and care. It enables the transfer of diagnostic and therapeutic innovations for the benefit of patients. The rapid involvement of industry in the transfer process and the interconnection between researchers and clinicians accelerate the development of new treatments.

3.64. Tumor biobank

This infrastructure cryopreserves tumor samples from cancer patients. Tumor biobanks respond to the need for genomic classification of tumors. By providing samples and associated medical data, tumor biobanks allow researchers to develop large-scale transcriptomic, proteomic or metabolomic analyses, making it possible to identify new prognostic or diagnostic *biomarkers*. The identification of these biomarkers has led to the development of *companion tests* for the use of targeted therapies.

Conclusion

Biobanks: What Perspectives and What Evolution?

The "professionalization" of biobanks has occurred very gradually over the last few years, but it has led to the establishment of very rigorous practices that harmonize the use of biological resources for research purposes. It is certainly thanks to the collective work carried out by actors from various professional backgrounds (biologists, pathologists, physicians, researchers, epidemiologists, bioinformaticians, hygienists, quality specialists, jurists, mathematicians, etc.) that the optimization of collections and the use of these biological resources have been able to develop over the years. This work has given rise to new quality standards and, in addition to the ISO 9001 certification and, more specifically for biobanking, the S96-900 certification established in France, a new *international standard for biobanks* has emerged, the ISO 20387 accreditation [MOU 20].

The optimization of biobanking work has also led to success stories in the scientific world, particularly in the medical world. Indeed, the quality and the good management of biological resources as well as *the harmonization of practices* from one biobank to another has made it possible to obtain reliable and therefore reproducible results. In addition to publications in often prestigious journals, these quality results can help in the development of diagnostic or prognostic biomarkers for a disease, or predictive ones for a therapeutic response, which can be transferred to daily practice and benefit patients. Quality biobanks also save time and money for the pharmaceutical and biotechnology industry, since the validation of certain therapeutic molecules or diagnostic tests requires rapid implementation based on the reproducibility of the results, in particular therapeutic results [FRE 15].

One of the turning points of biobanks is today, and will be even more so tomorrow, the integration of the considerable data associated with the sample and the patients, their families and the environment. The study of cohorts of several thousand or hundreds of thousands of individuals is one of the future challenges for biobanks. In fact, the biobank notion can give way to that of a clinical–biological database, thus associating the data specific to the samples and the data from the individuals and their evolution. It is certain that the mass of accumulated data will have to be exploitable and the intervention and training of new actors, such as "mathematician biologists", with a double competence must be anticipated from now on.

The amount of information generated from a biological sample, for example from whole genome sequencing, requires extraordinary storage capacities and material resources, which can only increase. Beyond the onerous task of maintaining and *securing information*, the obligations of confidentiality and of controlling access to personal data are also essential issues to be mastered from now on. One of the points that must be decided is whether the data associated with samples should be conserved, in particular nucleic acids.

The evolution of technologies means that the methods of analysis of genetic, epigenetic, transcriptomic, proteic and metabolomic parameters increasingly allow access to new information and generate data not previously acquired. In this context, is it really necessary to accumulate Big Data that risk being outdated and therefore becoming useless to exploit, or is it better to keep this information accessible in nucleic acids already extracted and preserved in a secure way, in order to be able to exploit them with new technologies to access more complete data later? Thus, methods for preserving nucleic acids at room temperature that require less storage space while also using ecosystem-protective approaches are proving to be particularly attractive [WAS 18, WAS 19].

One of the future challenges is also to reinforce health and safety measures, even if these measures have already been well-established in the past. The occurrence of new epidemics, or even new pandemics, requires increased vigilance from the different actors working in biobanks, especially when managing collections of biological products stored at low temperature, which have therefore not been fixed [HOF 20, HOF 21].

The future of biobanks probably involves two strategic priorities. The first is to reduce the collections of samples and to target collections towards certain samples of particular interest whose exploitation will reinforce and enrich the existing databases. The second is to maintain the capacity to react quickly to the emergence of new diseases [PEE 20]. It is only through the analysis of biological and tissue samples, correlated with clinical information, that the pathophysiological knowledge of diseases progresses and new treatments can be established. These actions also have several requirements, such as the *cooperation of biobanks*, within the same country as well as those existing in different countries and continents [LIG 21]. This is important in order to be able to freely exchange biological and clinical data and to give rise to *scientific cooperation* targeting, for example, rare or emerging diseases. The exchange of information also requires a common language and secure and compatible computer tools. Biobank software must therefore become more flexible and adapt to the different needs of users.

The sustainability of biobanks is essential to continue to increase scientific knowledge, and the business model of these biobanks must be controlled and monitored [RUS 20, SQA 20]. However, sustainability is linked to their recognition, which requires their individualization in order to become essential tools, and also to various communication actions that can make them more visible in their indispensable mission for the world of medical and scientific research.

Glossary

Accreditation: the formal recognition of technical and organizational skills, which validates a biobank's performance.

Agence nationale pour la sécurité du médicament et des produits de la santé **(ANSM):** the French National Agency for the Safety of Medicines and Health Products, a French public institution whose mission is to guarantee the safety of health products throughout their life cycle, during clinical trials and after marketing authorization.

Anonymization: the patient's last name, first name and social security number are encrypted. This coding facilitates the longitudinal follow-up of the patient, which can extend over several years during an epidemiological study.

Artificial intelligence: artificial intelligence allows quantitative evaluation of radiographic features through automatic processing of medical images, improving cancer screening, for example.

BBMRI-ERIC Biobank of Excellence or Expert Center: high-quality Expert Centers with medical and scientific expertise. They interact with academic centers of public research and private companies for medical and pharmaceutical research.

Biobanking and Biomolecular Resources Research Infrastructure (BBMRI): a network of biobanks that identifies sample collections, standardizes biobank operating procedures and shares samples and their associated data with the scientific community.

Biobanking and Biomolecular Resources Research Infrastructure–European Resource Infrastructure Consortium (BBMRI-ERIC): an infrastructure that brings together all biobanking stakeholders – researchers, biobankers, industry actors and patients – to accelerate medical research.

Biobank information management system (BIMS): the BIMS software ensures the traceability of samples and centralizes all data associated with a sample.

Biobank or biological resource center (BRC): these centers provide management of biological samples – storage, preservation, processing and distribution – as well as management of the associated data. Biobanks can manage human, animal, plant and microbial samples.

Biomarker: a quantitative indicator of a biological process – molecular, histological, radiographic or physiological – or of a therapeutic response.

Biospecimen: a sample of biological material (urine, blood, tissue, cells) from a human, animal, plant or microorganism.

Buffy coat: this fraction enriched in leukocytes and platelets is obtained after centrifugation of whole blood. This blood derivative is stored in a biobank for medical research and used in blood transfusion to produce platelet pools.

Certification: a certificate for products, processes or systems issued by a control body such as AFNOR for the French national standard or the ISO international standard.

Clinical biobank: clinical biobanks focus on given pathologies in order to understand their molecular and cellular mechanisms and discover therapeutic avenues.

Clinical research: clinical reasearch is organized into "clinical trials" in order to test the safety of the drug (phase I), the therapeutic efficacy of the product (phase II), the benefit/risk ratio (phase III) and, finally, post-marketing pharmacovigilance (phase IV).

Clinical study sponsor: the "sponsor" is an individual or legal entity that takes the initiative for medical research. They are responsible for setting up, managing and financing the clinical study and then entrusting it to an experienced "investigator" physician, a natural person who supervises the project.

Cohort and megacohorts: multifactorial studies include hundreds of thousands of participants to detect the onset of a disease. These epidemiological cohorts follow participants for several years.

Collection of biological samples: the constitution of a collection of human biological samples for research purposes is regulated by the French Public Health Code. As soon as the collection project is declared, the competent administrative authority checks that the conditions of constitution, conservation and exploitation of the collection ensure the proper use, security and confidentiality of the data collected.

***Comité de protection des personnes* (CPP):** the French Committee for the Protection of Individuals. In order to protect participants in biomedical research, law no. 2004-806 on public health, known as the "Huriet–Sérusclat" law, requires that clinical research protocols be submitted to an advisory committee for the protection of individuals. The favorable opinion of the CPP is mandatory to start a research project.

***Commission nationale de l'informatique et des libertés* (CNIL):** the French National Commission for Information Technology and Civil Liberties. In French law, the protection of personal data is legally regulated by the law on information technology and civil liberties (1978), which the CNIL is responsible for enforcing.

Companion test: a diagnostic test to select patients who are likely to respond favorably to a treatment.

Computed tomography (CT): a medical imaging test that is commonly used for cancer screening and for follow-up of patients, as well as to evaluate structures.

Diagnostic biomarker: molecular or in combination with medical imaging, diagnostic biomarkers are used to identify carriers of disease.

Digital health or e-health: artificial intelligence is capable of making early diagnoses (preventive medicine), screening billions of molecules and identifying biomarkers of a disease, offering new therapeutic leads.

Electronic case report form (eCRF): all individual information collected during the clinical study – patient identification, study inclusion criteria, informed consent and all medical test results.

European, Middle Eastern and African Society for Biopreservation and Biobanking (ESBB): this network of biobanks shares biological resources in the fields of health, education and the environment between countries in the Europe, Middle East and Africa (EMEA) region.

Food and Drug Administration (FDA): this service of the US administration controls food and pharmaceutical products.

Formalin-fixed, paraffin-embedded (FFPE) tissue: dehydrated, formalin-fixed tissue samples are embedded in paraffin. Stored at room temperature in the form of solid blocks, they can be easily cut into thin sections with a microtome, placed on histology slides and then stained.

Free and informed consent: by signing a consent form, an individual expresses willingness to participate in a research project. This consent, transmitted to the CPP, is mandatory when it comes to the use of biological samples and personal data.

General Data Protection Regulation (GDPR): this European regulation protects the confidentiality of personal data and applies to the activities of biobanks concerning the collection of personal data and the creation of new data from and associated with samples.

Imaging biobank: a repository of radiology imaging data and biomarkers, shared among researchers and linked to biomedical banks.

Imaging biomarker: by measuring a volume or a diameter using medical imaging, these biomarkers allow the detection of early signs, even before the appearance of clinical signs.

International Society for Biological and Environmental Repositories (ISBER): this network of biobanks is dedicated to preserving and conserving collections of plants, seeds, animal organisms and environmental microorganisms.

ISO 20387:2018 Biotechnologies – Biobanking – general requirements for biobanking: this ISO international standard is specific to the activities of biobanks; it describes the requirements for competence, impartiality and consistent operation in biobanks.

Laboratory information management system: this system oversees a sample's journey through the biobank from the moment it is received, through to preparation, storage and, finally, the distribution of the sample and associated data.

Machine learning and deep learning: deep learning, a subcategory of machine learning, is the most developed form of artificial intelligence used in medical imaging. The algorithm performs diagnostic tasks, the most common of which are detection, object segmentation and automated extraction of imaging descriptors.

Material transfer agreement (MTA): an agreement by which a biobank transfers biological material to an academic or industrial partner.

Medical imaging: clinical image acquisition techniques such as computed tomography, positron emission tomography and magnetic resonance imaging have made it possible to visualize organs and the entire body.

Monitoring biomarker: routinely measured, these biomarkers assess the status of a disease or exposure to a medical or environmental product.

NF S96-900: this French standard specific to biobanks provides requirements for the quality management systems of BRCs in order to facilitate the acquisition, validation, conservation and distribution of biological resources by guaranteeing the quality and traceability of samples.

Omics: omics technologies include genomic (DNA sequencing), transcriptomic (all transcribed messenger RNA), proteomic (analysis of proteins in the sample) and metabolomic (small molecules or metabolites present in the sample) analyses.

Peripheral blood mononuclear cells (PBMCs): gathering of mononuclear leukocytes, monocytes and T, B and NK lymphocytes from adult peripheral blood or fetal cord blood.

Personalized or individualized medicine: personalized medicine has transformed the one-size-fits-all healthcare offer into an individual approach by using omics technologies for genetic, proteomic, transcriptomic and metabolomic analyses.

Pharmacodynamic/response biomarker: a biochemical, histological and anatomical parameter that assesses the response to a treatment.

Population biobanking: the collection of samples from non-selected volunteers from the general population over several years. This prospective approach is the strength of medical prevention programs.

Positron emission tomography (PET): this medical imaging test is based on the injection of a weakly radioactive tracer into blood, which provides functional information on the metabolism and proliferation of a tumor.

Pre-analytical and analytical phase: pre-analytical variables, such as the nature of the sample and the storage technique and its duration, are part of the post-sampling process, the control of which ensures the integrity and therefore the quality of the sample.

Precision medicine: precision medicine provides a therapeutic lead based on the genetic (omics technologies) and environmental (microbiome and exposome) profile of each individual.

Predictive biomarker: these biomakers target specific groups of people for whom a particular drug may be useful. The diagnostic test associated with the treatment is called the "companion" test.

Prognostic biomarker: these biomarkers identify the probability of a clinical event, such as recurrence or progression of a disease, independent of drug therapy.

Prospective survey/study: a prospective study takes place over several decades, during which the participants fill in health questionnaires and describe their lifestyle. Regularly, they undergo complete medical check-ups.

QR code: a 2D data matrix code that gathers the minimum information about patients – identification number, age, gender and primary diagnosis.

Quality management system (QMS): the quality strategy of a biobank, which must meet the objectives of quality assurance and quality control. An effective QMS is necessary to obtain accreditation for a biobank and thus gain recognition from the scientific community.

Radiomics: the automatic extraction of quantitative parameters (descriptors) from a medical image. This information is related to the patient's phenotype and can be correlated with the gene expression profile (radiogenomics).

Region of interest (ROI) or volume of interest (VOI): a manually or automatically delimited area in medical imaging that measures the volume, shape and texture of a tumor.

Retrospective survey/study: the discovery of new candidate biomarkers leads to retrospective studies, enabled by large sample collections and accompanied by epidemiological, clinical, histological and genomic data.

Standard operating procedures (SOPs): a high level of standardization implies the use of state-of-the-art technologies, circular assay evaluations and harmonized operating procedures – within a biobank or biobank network – for pre-analytical steps, molecular analysis and complex data management.

Stratified medicine: groups of patients are "stratified" according to their genetic profile in order to individualize therapy and deliver tailored treatment to patients who are likely to respond to a treatment.

Translational research: a link between academic research and clinical research that enables the transfer of diagnostic and therapeutic innovations for the benefit of patients.

Tumor biobank: this infrastructure (biobank) cryopreserves tumor samples from cancer patients.

References

Foreword

[CLÉ 19] CLÉMENT B., DUYCKAERTS C., HAUW J.J. et al., "Pérenniser les centres de ressources biologiques : un enjeu majeur pour la recherche biomédicale", *Bull. Acad. Natl. Med.*, no. 203, pp. 2–8, 2019.

Introduction

[ALD 19] AL DIFFALHA S., SEXTON K.C., WATSON P.H. et al., "The importance of human tissue bioresources in advancing biomedical research", *Biopreserv. Biobank.*, vol. 17, pp. 209–212, 2019.

[BAI 16] BAIRD P.M., GUNTER E.W., VAUGHT J., "Building a biobank", *Biopreserv. Biobank.*, no. 14, pp. 87–88, 2016.

[BAR 20] BARNES R.O., WATSON P.H., "Precision medicine: Driving the evolution of biobanking quality", *Healthc. Manage. Forum*, vol. 33, pp. 102–106, 2020.

[CLÉ 14] CLÉMENT B., YUILLE M., ZATLOUKAL K. et al., "EU–US expert group on cost recovery in biobanks, Public biobanks: Calculation and recovery of costs", *Sci. Transl. Med.*, no. 6, pp. 261–345, 2014.

[DOU 17] DOUCET M., BECKER K.F., BJORKMAN J. et al., "Quality matters: 2016 annual conference of the national infrastructures for biobanking", *Biopreserv. Biobank.*, vol. 195, no. 10, 2017.

[ELL 15] ELLERVIK C., VAUGHT J., "Preanalytical variables affecting the integrity of human biospecimens in biobanking", *Clin. Chem.*, no. 61, pp. 914–34, 2015.

[HAR 19] HARTMAN V., MATZKE L., WATSON P.H., "Biospecimen complexity and the evolution of biobanks", *Biopreserv. Biobank.*, no. 17, pp. 264–270, 2019.

[HEW 11] HEWITT R.E., "Biobanking: The foundation of personalized medicine", *Curr. Opin. Oncol.*, vol. 23, pp. 112–119, 2011.

[HOF 13] HOFMAN V., ILIE M., LONG E. et al., "Measuring the contribution of tumor biobanks to research in oncology: Surrogate indicators and bibliographic output", *Biopreserv. Biobank.*, no. 11, pp. 235–44, 2013.

[HOF 14] HOFMAN P., BRÉCHOT C., ZATLOUKAL K. et al., "Public–private relationships in biobanking: A still underestimated key component of open innovation", *Virchows. Arch.*, no. 464, pp. 3–9, 2014.

[HOF 20] HOFMAN P., PUCHOIS P., BREST P. et al., "Possible consequences of the COVID-19 pandemic on the use of biospecimens from cancer biobanks for research in academia and bioindustry", *Nat. Med.*, no. 260, p. 8, 2020.

[HUA 20] HUANG S.F., HUANG Y.C., CHANG F.Y. et al., "Rapid establishment of a COVID-19 biobank in NHRI by National Biobank Consortium of Taiwan", *Biomed. J.*, no. 43, pp. 314–317, 2020.

[KIN 20] KINKOROVA J., TOPOLCAN O., "Biobanks in the era of big data: Objectives, challenges, perspectives, and innovations for predictive, preventive, and personalised medicine", *EPMA J.*, no. 11, pp. 333–341, 2020.

[LIN 20] LIN J.C., HSIAO W.W., FAN C.T., "Transformation of the Taiwan Biobank 3.0: Vertical and horizontal integration", *J. Transl. Med.*, no. 18, vol. 304, 2020.

[MAR 20] MARINO S., ZHAO Y., ZHOU N. et al., "Compressive big data analytics: An ensemble meta-algorithm for high-dimensional multisource datasets", *PLoS One*, 2020.

[NAR 20] NARITA A., UEKI M., TAMIYA G., "Artificial intelligence powered statistical genetics in biobanks", *J. Hum. Genet.*, 2020.

[PEE 20] PEELING R.W., BOERAS D., WILDER-SMITH A. et al., "Need for sustainable biobanking networks for COVID-19 and other diseases of epidemic potential", *Lancet. Infect. Dis.*, no. 20, pp. 268–273, 2020.

[RAO 19] RAO A., VAUGHT J., TULSKIE B. et al., "Critical financial challenges for biobanking – Report of a National Cancer Institute study", *Biopreserv. Biobank.*, no. 17, pp. 129–138, 2019.

[ROV 20] ROVERE-QUERINI P., TRESOLDI C., CONTE C. et al., "Biobanking for COVID-19 research", *Panminerva Med.*, 19 October 2020.

[RUS 19] RUSH A., MATZKE L., COOPER S. et al., "Research perspective on utilizing and valuing tumor biobanks", *Biopreserv. Biobank.*, vol. 17, pp. 219–229, 2019.

[SIM 14] SIMEON-DUBACH D., WATSON, P.P., "Biobanking 3.0: Evidence based and customer focused biobanking", *Clin. Biochem.*, no. 47, p. 300, 2014.

[SIM 20] SIMEON DUBACH D., ROEHRL M.H., HOFMAN P. et al., "Enhancing cooperation between academic biobanks and biomedical industry: Better mutual understanding and new collaborative models are needed", *Biopreserv. Biobank.*, no. 18, pp. 144–149, 2020.

[TAN 20] TANGA V., LEROY S., FAVADA J. et al., "Establishment of a collection of blood-derived products from COVID-19 patients for translational research: Experience of the LPCE Biobank (Nice, France)", *Biopreserv. Biobank.*, 15 December 2020.

[VAU 11] VAUGHT J., ROGERS J., CAROLIN T. et al., "Biobankonomics: Developing a sustainable business model approach for the formation of a human tissue biobank", *J. Natl. Cancer Inst. Monogr.*, pp. 24–31, 2011.

[VAU 14] VAUGHT J., BLEDSOE M., WATSON P., "Biobanking on multiple continents: Will international coordination follow?", *Biopreserv. Biobank.*, no. 12, pp. 1–2, 2014.

[VAU 19] VAUGHT J., "Biobanking and beyond: The importance of scientific collections", *Biopreserv. Biobank.*, vol. 17, no. 1, p. 1, 2019.

[VAU 20] VAUGHT J., "Biobanking during the COVID-19 pandemic", *Biopreserv. Biobank.*, no. 18, pp. 153–154, 2020.

[WAT 17] WATSON P.H., "Biospecimen complexity – The next challenge for cancer research biobanks?", *Clin. Cancer Res.*, no. 23, pp. 894–898, 2017.

[WAT 19] WATSON P.H., HEWITT R.E., CATCHPOOLE D.R. et al., "Biobank: What's in a name?", *Biopreserv. Biobank.*, vol. 17, pp. 204–208, 2019.

Chapter 1

[AIR 19] AIRAGNES G., LEMOGNE C., RENUY A. et al., "Prevalence of prescribed benzodiazepine long-term use in the French general population according to sociodemographic and clinical factors: Findings from the CONSTANCES cohort", *BMC Public Health*, vol. 19, pp. 566–575, 2019.

[AMA 20] AMALFITANO M., DONG Q., KEITA A., "Impact of the pre-analytical phase on the quality of samples collected in biobanks", *Med. Sci.*, vol. 36, pp. 277–279, 2020.

[BAI 05] BAIRD P., FROME R.J., "Large-scale repository design", *Cell Preserv. Technol.*, vol. 3, pp. 256–266, 2005.

[BAT 14] BATEMAN S., "Personalized medecine: An elusive concept, diversified practices", *Med. Sci.*, vol. 30, no. 2, pp. 8–13, 2014.

[BBM 21] BBMRI-ERIC, "BBMRI-ERIC", available at: https://directory.bbmri-eric.eu/menu/main/app-molgenis-app-biobank-explorer/biobankexplorer, 2021.

[BET 18] BETSOU F., BILBAO R., CASE J. et al., "Standard PREanalytical Code version 3.0", *Biopreserv. Biobank.*, 2018.

[BGI 21] BGI, "BGI develops real-time fluorescent RT-PCR kit for detecting the 2019 novel coronavirus", available at: https://www.bgi.com/global/company/news/bgi-develops-real-time-dna-based-kit-for-detecting-the-2019-novel-coronavirus/, 2021.

[BIO 21] BIOBANK GRAZ, "BIOBANK GRAZ", available at: https://biobank.medunigraz.at/en/, 2021.

[BRA 14] BRADBURY K.E., APPLEBY P.N., KEY T.J., "Fruit, vegetable, and fiber intake in relation to cancer risk: Findings from the European Prospective Investigation into Cancer and Nutrition (EPIC)", *Am. J. Clin. Nutr.*, vol. 1, pp. 394–398, 2014.

[BYC 18] BYCROFT C., FREEMAN C., PETKOVA D. et al., "The UK Biobank resource with deep phenotyping and genomic data", *Nature*, vol. 562, pp. 203–209, 2018.

[CHA 16] CHASSANG G., CAMBON THOMSEN A., RIAL-SEBBAG E. et al., *Éthique et réglementation des biobanques de recherche. Aspects opérationnels et perspectives*, Biobanques, Paris, 2016.

[CHE 20] CHEN F.Z., YOU L.J., YANG F. et al., "CNGBdb: China National GeneBank DataBase", *Yi Chuan*, vol. 42, no. 8, pp. 799–809, 20 August 2020.

[CHU 98] CHU J.Y., HUANG W., KUANG S.Q. et al., "Genetic relationship of populations in China", *Proc. Natl. Acad. Sci. USA*, vol. 95, pp. 11763–11768, 1998.

[CLA 15] CLAVEL-CHAPELON F., E3N STUDY GROUP, "Cohort profile: The French E3N Cohort Study", *Int. J. Epidemiol.*, vol. 44, pp. 801–809, 2015.

[COL 16] COLLINS F.S., "Preparing to launch the precision medicine initiative cohort program", available at: https://www.nih.gov/about-nih/who-we-are/nih-director/statements/preparing-launch-precision-medicine-initiative-cohort-program, 2016.

[CYR 10] CYRANOSKI D., "Chinese bioscience: The sequence factory", *Nature*, vol. 464, pp. 22–24, 2010.

[DAW 51] DAWBER T.R., MEADORS G.F., MOORE F.F., "Epidemiologic approach to heart disease: The Framingham study", *Am. J. Public Health*, vol. 41, pp. 279–286, 1951.

[DAW 66] DAWBER T.R., KANNEL W.B., "The Framingham study, an epidemiologic approach to coronary heart diseases", *Circulation*, vol. 34, pp. 553–555, 1966.

[DEN 19] DENNY J.C., RUTTER J.L., GOLDSTEIN D.B. et al., "The 'All of Us' research program", *N. Engl. J. Med.*, vol. 381, pp. 668–676, 2019.

[ELL 18] ELLIOT L.T., SHARP K., ALFARO-ALMAGRO F. et al., "Genome-wide association studies of brain imaging phenotypes in UK Biobank", *Nature*, vol. 562, pp. 210–216, 2018.

[EUR 16] EUROPEAN UNION, Regulation (EU) 2016/679 of the European Parliament and of the Council, OJ L 119 04.05.2016, p. 1, 2016.

[FDA 16] FDA, "Use of electronic informed consent in clinical investigations – Questions and answers, Guidance for institutional review boards, investigators, and sponsors", available at: http://www.fda.gov/downloads/drugs/guidance complianceregulatoryinformation/guidances/ucm436811.pdf, 2016.

[FRA 15] FRANSSON M.N., RIAL-SEBBAG E., BROCHHAUSEN M. et al., "Toward a common language for biobanking", *Eur. J. Hum. Genet.*, no. 23, pp. 22–28, 2015.

[FRE 04a] FRENCH GOVERNMENT, Loi no. 2004-806/loi Huriet-Sérusclat, *JO*, 2004.

[FRE 04b] FRENCH GOVERNMENT, Loi no. 2004-801/loi CNIL, *JO*, 2004.

[FRE 12] FRENCH GOVERNMENT, Loi no. 2012-300/loi Jardé, *JO*, 2012.

[FRE 15] FREEDMAN L.P., COCKBURN I.M., SIMCOE T.S., "The economics of reproducibility in preclinical research", *PLoS Biol.*, vol. 13, pp. 1–9, 2015.

[GAN 15] GAN R., WANG H., SONG Y. et al., "Chinese biobanking initiatives", *Biopreserv. Biobank.*, vol. 13, pp. 4–7, 2015.

[GOD 04] GODARD B., MARSHALL J., LABERGE C. et al., "Strategies for consulting with the community: The cases of four large-scale genetic databases", *Sci. Eng. Ethics.*, vol. 10, pp. 457–477, 2004.

[HEW 13] HEWITT R., WATSON P., "Defining biobank", *Biopreserv. Biobank.*, vol. 11, pp. 309–315, 2013.

[HOL 16] HOLUB P., SWERTZ M., REIHS R. et al., "BBMRI-ERIC directory: 515 biobanks with over 60 million biological samples", *Biopreserv. Biobank.*, vol. 14, pp. 559–562, 2016.

[HUA 09] HUANG S., LI R., ZHANG Z. et al., "The genome of the cucumber Cucumis sativus L.", *Nat. Genet.*, vol. 41, pp. 1275–1281, 2009.

[HUP 16] HUPPERTZ B., BAYER M., MACHEINER T. et al., "Biobank Graz: The hub for innovative biomedical research", *Open J. Bioresour.*, vol. 3, no. 3, 2016.

[INT 05] INTERNATIONAL CHICKEN GENOME SEQUENCING CONSORTIUM, "Sequence and comparative study analysis of the chicken genome provide unique perspectives on vertebrate evolution", *Nature*, vol. 432, pp. 695–716, 2004.

[LAZ 14] LAZARIDIS K.N., MCALLISTER T.M., BADOVIC-VUKSANOVIC D. et al., "Implementing individualized medecine into the medical practice", *Am. J. Med. Genet. Part C Semin. Med. Genet.*, vol. 166, pp. 15–23, 2014.

[LEH 12] LEHMANN S., GUADAGNI F., MOORE H. et al., "Impact standard preanalytical coding for biospecimens: Review and implementation of the Sample PREanalytical Code (SPREC)", *Biopreserv. Biobank.*, vol. 10, pp. 366–374, 2012.

[LEH 16] LE HIR P., "Santé : un français sur deux est en surpoids", *Le Monde*, available at: https://www.lemonde.fr/planete/article/2016/10/25/un-francais-sur-deux-est-en-surpoids_5019615_3244.html, 2016.

[LEI 15] LEITSALU L., HALLER T., ESKO T. et al., "Cohorte profile: Estonian biobank of the Estonian Genome Center, University of Tartu", *Int. J. Epidemiol.*, vol. 44, pp. 1137–1147, 2015.

[LI 10] LI R., FAN W., TIAN G. et al., "The sequence and de novo assembly of the giant panda genome", *Nature*, vol. 463, pp. 311–317, 2010.

[LIN 19] LINSEN L., T'JOEN V., VAN DER STRAETEN C. et al., "Biobank quality management in the BBMRI.be network", *Front. Med.*, vol. 6, pp. 141–150, 2019.

[LU 20] LU R., ZHAO X., LI J. et al., "Genomic characterisation and epidemiology of 2019 novel coronavirus: Implications for virus origins and receptor binding", *Lancet*, vol. 395, pp. 565–574, 2020.

[MAT 16] MATTA J., ZINS M., FERAL-PIERSSENS A.L. et al., "Prévalence du surpoids, de l'obésité et des facteurs de risque cardio-métaboliques dans la cohorte Constances", *Bull. epidemiol. Hebd. J. Med.*, vol. 35–36, pp. 640–646, available at: http://invs.santepubliquefrance.fr/beh/2016/35-36/2016_35-36_5.html, 2016.

[MOO 11] MOORE H.M., KELLY A., JEWEL S.D. et al., "Biospecimen reporting for improved study quality", *Biopreserv. Biobank.*, vol. 9, pp. 57–70, 2011.

[OEC 01] OECD, Les centres de ressources biologiques – Fondements du développement des sciences de la vie et des biotechnologies, Report, OECD, Paris, 2001.

[QIN 03] QIN E., ZHU Q., YU M. et al., "A complete sequence and comparative analysis of a SARS-associated virus (Isolate BJ01)", *Chin. Sci. Bull.*, vol. 48, pp. 941–948, 2003.

[RAS 10] RASMUSSEN M., LI Y., LINDGREEN S. et al., "Ancient human genome sequence of an extinct paleo-eskimo", *Nature*, vol. 463, pp. 757–762, 2010.

[SAN 17] SANKAR P., PARKER L., "The precision medicine initiative's All of Us research program: An agenda for research on its ethical, legal and social issues", *Genet. Med.*, vol. 19, pp. 743–750, 2017.

[STO 16] STOEKLE H.C., MAMZER-BRUNEEL M.F., VOGT G. et al., "23andMe: A new two-sided data-banking market model", *BMC Med. Ethics*, vol. 17, pp. 19–30, 2016.

[TRI 14] TRICHOPOULOU A., ORFANOS P., NORAT T. et al., "Modified Mediterranean diet and survival: EPIC-elderly prospective cohort study", *BMJ*, vol. 330, pp. 991–998, 2014.

[VAN 19] VAN DER VELDE K.J., IMHANN F., CHARBON B. et al., "MOLGENIS research: Advanced bioinformatics data software for non-bioinformaticians", *Bioinformatics*, vol. 35, pp. 1076–1078, 2019.

[WAS 18] WASHETINE K., HEEKE S., BONNETAUD C. et al., "Establishing a dedicated lung cancer biobank at the University Center Hospital of Nice (France). Why and how?", *Cancers*, vol. 10, pp. 220–235, 2018.

[WOJ 18] WOJCICKI A., "A note on 23andMe's new collaboration with GSK", available at: https://blog.23andme.com/news/collaboration-with-gsk/, 2018.

[XIA 04] XIA Q., ZHOU Z., LU C. et al., "A draft sequence for the genome of the domesticated silkworm (*Bombix mori*)", *Science*, vol. 306, pp. 1937–1940, 2004.

[YU 02] YU J., HU S., WANG J. et al., "A draft sequence of the rice genome (*Oryza sativa* L. ssp *indica*)", *Science*, vol. 296, pp. 79–92, 2002.

Chapter 2

[AER 14] AERTS H., VELAZQUEZ E., LEIJENAAR R. et al., "Decoding tumour phenotype by noninvasive imaging using a quantitative radiomics approach", *Nat. Commun.*, vol. 5, pp. 1–8, 2014.

[AER 18] AERTS, H., "Data science in radiology: A path forward", *Clin. Cancer Res.*, vol. 24, pp. 532–534, 2018.

[ALZ 20] ALZHEIMER'S DISEASE NEUROIMAGING INITIATIVE, "Alzheimer's Disease Neuroimaging Initiative", available at: http://adni.loni.usc.edu/, 2020.

[ARM 11] ARMATO III S.G., MCLENNAN G., BIDAUT L. et al., "The Lung Image Database Consortium (LIDC) and Image Database Resource Initiative (IDRI): A completed reference database of lung nodules on CT scans", *Med. Phys.*, vol. 38, pp. 915–931, 2011.

[AUF 16] AUFFRAY C., BALLING R., BARROSO I. et al., "Making sense of big data in health research: Towards an EU action plan", *Genome Med.*, vol. 8, pp. 71–84, 2016.

[AYA 18] AYACHE N., "L'imagerie médicale à l'heure de l'intelligence artificielle", in VILLANI C. and NORDLINGER B. (eds), *Santé et intelligence artificielle*, CNRS Éditions, Paris, 2018.

[BAM 15] BAMBERG F., KAUCZOR H.U., WECKBACH S. et al., "Whole-body MR imaging in the german national cohort: Rationale, design, and technical background", *Radiology*, vol. 277, pp. 206–220, 2015.

[BBM 16] BBMRI-ERIC, "A big step towards personalised medecine: CBmed appointed first Expert Centre of European biobanks", available at: https://www.bbmri-eric.eu/news-events/cbmed/, 2016.

[BEG 12] BEGLEY, C.G., ELLIS, L.M., "Drug development: Raise standards for preclinical cancer research", *Nature*, vol. 483, pp. 531–533, 2012.

[BRE 94] BRETELER M.M., VAN SWIETEN J.C., BOTS M.L. et al., "Cerebral white matter lesions, vascular risk factors, and cognitive function in a population-based study: The Rotterdam study", *Neurology*, vol. 44, pp. 1246–1252, 1994.

[BRU 15] BRUNO M.A., WALKER E.A., ABUJUDEH H.H., "Understanding and confronting our mistakes: The epidemiology of error in radiology and strategies for error reduction", *Radiographics*, vol. 35, pp. 1668–1676, 2015.

[BRU 19] BRUNELLE F., BRUNELLE P., "Artificial intelligence and medical imaging: Definition, state of the art and perspectives", *Bull. Acad. Nat. Med.*, vol. 203, pp. 683–687, 2019.

[BUH 14] BUHL T., LEGLER T.J., ROSENBERGER A. et al., "Controlled-rate freezer cryopreservation of highly concentrated peripheral blood mononuclear cells results in higher cell yields and superior autologous T-cell stimulation for dendritic cell-based immunotherapy", *Cancer Immunol. Immunother.*, vol. 61, pp. 2021–2031, 2014.

[CAL 18] CALIFF R.M., "Biomarker definitions and their applications", *Exp. Biol. Med.*, vol. 243, pp. 213–221, 2018.

[CAR 15] CAROTENUTO D., LUCHINAT C., MARCON G. et al., "The Da Vinci European biobank: A metabolomics-driven infrastructure", *J. Pers. Med.*, vol. 5, pp. 107–119, 2015.

[CNA 16] CNAG, "Certification for quality of CNAG genomic data", available at: https://elixir-europe.org/news/certification-quality-cnag-genomic-data, 2016.

[DAN 19] DANAT I.M., CLIFFORD A., PARTRIDGE M. et al., "Impacts of overweight and obesity in older age on the risk of dementia: A systematic literature review and a meta-analysis", *J. Alzheimers Dis.*, vol. 70, pp. 87–99, 2019.

[DEK 19] DEKKERS I.A., JANSSEN P.R., LAMB H.J., "Obesity, brain volume, and white matter microstructure at MRI: A cross-sectional UK Biobank study", *Radiology*, vol. 291, pp. 763–771, 2019.

[DEM 19] DEMICHELIS R., "Therapixel lève 5 millions et met l'IA au service du diagnostic des cancers du sein", *Les Echos Business*, available at: https://business.lesechos.fr/entrepreneurs/financer-sa-creation/0600861942175-therapixel-leve-5-millions-pour-mettre-l-ia-au-service-du-diagnostic-des-cancers-du-sein-327716.php, 2019.

[EAR 20] EARY J.F., "Cancer imaging program update: 2020", *Radiol. Imaging Cancer*, vol. 2, no. 4, e204021, 2020.

[EIS 09] EISENHAUER E.A., THERASSE P., BOGAERTS J. et al., "New response evaluation criteria in solid tumours: Revised RECIST guideline (version 1.1)", *Eur. J. Cancer*, vol. 45, pp. 228–247, 2009.

[EUR 15] EUROPEAN SOCIETY OF RADIOLOGY, "The ESR position paper on imaging biobanks", *Insights Imaging*, vol. 6, pp. 403–410, 2015.

[EUR 20a] EUROPEAN SOCIETY OF RADIOLOGY, "ESR Action Plan for Medical Imaging", available at: https://www.myesr.org/eu-international-affairs/esr-action-plan-medical-imaging, 2020.

[EUR 20b] EURO-BIOIMAGING, Projet européen Euro-Bioimaging, 2020.

[EVA 16] EVANS T.E., O'SULLIVAN M.J., DE GROOT M. et al., "White matter microstructure improves stroke risk prediction in the general population", *Stroke*, vol. 47, pp. 2756–2762, 2016.

[EXC 14] EXCEMET, "Expert Center for Metabolomics", available at: http://www.excemet.org/, 2014.

[FDA 16] FDA–NIH BIOMARKER WORKING GROUP, "Biomarkers, Endpoints, and Other Tools (BEST) Resource", available at: www.ncbi.nlm.nih.gov/books/NBK326791/, 2016.

[HAN 14] HANTZSCH M., TOLIOS A., BEUTNER F. et al., "Comparison of whole blood RNA preservation tubes and novel generation RNA extraction kits for analysis of mRNA and MiRNA profiles", *PLoS One*, vol. 9, pp. 1–17, 2014.

[HEN 13] HENNY J., ZINS M., GOLDBERG M. et al., *Guide pour la constitution d'une biobanque associée aux études épidémiologiques en population générale*, Lavoisier, Paris, 2013.

[IDA 20] IDA, "Collections of medical images from the ADNI image data archives", available at: https://ida.loni.usc.edu/login.jsp?project=ADNI, 2020.

[IKR 15] IKRAM M.A., VAN DER LUGT A., NIESSEN W.J. et al., "The Rotterdam scan study: Design update 2016 and main findings", *Eur. J. Epidemiol.*, vol. 30, pp. 1299–1315, 2015.

[IKR 17] IKRAM M.A., BRUSSELLE G.G.O., MURAD S.D. et al., "The Rotterdam study: 2018 update on objectives, design and main results", *Eur. J. Epidemiol.*, vol. 32, pp. 807–850, 2017.

[IKR 20] IKRAM M.A., BRUSSELLE G.G.O., GHANBARI M. et al., "Objectives, design and main findings until 2020 from the Rotterdam study", *Eur. J. Epidemiol.*, vol. 35, pp. 483–517, 2020.

[JAF 06] JAFFE C., "Measures of response: RECIST, WHO, and new alternatives", *J. Clin. Oncol.*, vol. 24, pp. 3245–3251, 2006.

[KOZ 19] KOZLAKIDIS Z., "The digital revolution of biobanks: What role is precision medicine playing?", *Health Europa*, available at: https://www.healtheuropa.eu/biobanks-precision-medicine/90687/, 2019.

[LAM 12] LAMBIN P., RIOS-VELAZQUEZ E., LEIJENAAR R. et al., "Radiomics: Extracting more information from medical images using advanced feature analysis", *Eur. J. Cancer*, vol. 48, pp. 441–446, 2012.

[LÉG 78] LÉGIFRANCE, Loi no. 78–17, 1978.

[LEQ 20] LE QUEAU C., PHILLIPS W.A., SRINIVASAN A., "Practical use of biobank information management system (BIMS) for clinical-biological annotations in biobanks", *Medecine/Sciences*, vol. 36, pp. 285–288, 2020.

[LIA 19] LIAO F., LIANG M., LI Z. et al., "Evaluate the malignancy of pulmonary nodules using the 3-D Deep Leaky Noisy-OR Network", *IEEE Transactions on Neural Networks and Learning Systems*, vol. 30, pp. 3484–3495, 2019.

[MCN 07] MCNITT-GRAY M.F., ARMATO III S.G., MEYER C.R. et al., "The Lung Image Database Consortium (LIDC) data collection process for nodule detection and annotation", *Acad. Radiol.*, vol. 14, pp. 1464–1474, 2007.

[MED 16] MEDICAL UNIVERSITY OF VIENNA, "New CBmed Core Lab opened at MedUni Vienna", available at: https://www.meduniwien.ac.at/web/en/about-us/news/detailsite/2016/news-from-july-2016/new-cbmed-core-lab-opened-at-meduni-vienna/, 2016.

[NAK 20] NAKO, "German National Cohort", available at: https://nako.de/, 2020.

[NAT 11] NATIONAL LUNG SCREENING TRIAL RESEARCH TEAM, ABERLE D.R., ADAMS A.M. et al., "Reduced lung-cancer mortality with low-dose computed tomographic screening", *N. Engl. J. Med.*, vol. 365, pp. 395–409, 2011.

[OLU 07] OLUND G., LINDQVIST P., LITTON J.E., "BIMS: An information management system for biobanking in the 21st century", *IBM Systems Journal*, vol. 46, pp. 171–182, 2007.

[ORC 18] ORCHARD-WEBB D., "10 top laboratory information management systems (LIMS) for biobanking", available at: https://www.biobanking.com/top-lims/, 2018.

[PRE 16] PRESS G., "Cleaning big data: Most time-consuming, least enjoyable data science task, survey says", *Forbes*, available at: https://www.forbes.com/sites/gilpress/2016/03/23/data-preparation-most-time-consuming-least-enjoyable-data-science-task-survey-says/#72f481636f63More, 2016.

[SCH 16] SCHOLL M., LOCKHART S.N., SCHONHAUT D.R. et al., "PET imaging of Tau deposition in the aging human brain", *Neuron.*, vol. 89, pp. 971–982, 2016.

[SCH 20] SCHAFFTER T., BUIST D.S.M., LEE C.I. et al., "Evaluation of combined artificial intelligence and radiologist assessment to interpret screening mammograms", *JAMA Netw. Open.*, vol. 3, no. 3, p. e200265, 2020.

[SET 17] SETIO A.A.A., TRAVERSO A., DE BEL T. et al., "Validation, comparison, and combination of algorithms for automatic detection of pulmonary nodules in computed tomography images: The LUNA16 challenge", *Med. Image Anal.*, vol. 42, pp. 1–13, 2017.

[SPA 11] SPATH M.B., GRIMSON J., "Applying the archetype approach to the database of a biobank information management system", *Int. J. Med. Inform.*, vol. 80, pp. 205–226, 2011.

[SPL 07] SPLANSKY G.L., COREY D., YANG Q. et al., "The third generation cohort of the National Heart, Lung, and Blood Institute's Framingham Heart Study: Design, recruitment, and initial examination", *Am. J. Epidemiol.*, vol. 165, pp. 1328–1335, 2007.

[STA 16] STANTA G., WENZEL JAHN S., BONIN S. et al., "Tumour heterogeneity: Principles and practical consequences", *Virchows. Archiv.*, vol. 469, pp. 371–384, 2016.

[UKB 20] UK BIOBANK, "UK Biobank imaging study", available at: https://imaging.ukbiobank.ac.uk/#introduction, 2020.

[VAN 11] VAN GINNEKEN B., ARMATO III S.G., DE HOOP B. et al., "Comparing and combining algorithms for computer-aided detection of pulmonary nodules in computed tomography scans: The ANODE09 study", *Med. Image Anal.*, vol. 14, pp. 707–722, 2011.

[VAN 15] VAN OMMEN G.-J.B., TÖRNWALL O., BRÉCHOT C. et al., "BBMRI-ERIC as a resource for pharmaceutical and life science industries: The development of biobank-based Expert Centres", *Eur. J. Hum. Genet.*, vol. 23, pp. 893–900, 2015.

[VIA 18] VIAL A., STIRLING D., FIELD M. et al., "The role of deep learning and radiomic feature extraction in cancer-specific predictive modelling: A review", *Transl. Cancer Res.*, vol. 7, pp. 803–816, 2018.

[WEI 17] WEINER M.W., VEITCH D.P., AISEN P.S. et al., "The Alzheimer's Disease Neuroimaging Initiative 3: Continued innovation for clinical trial improvement", *Alzheimer Dement.*, vol. 13, pp. 561–571, 2017.

[WIL 16] WILKINSON M.D., DUMONTIER M., AALBERSBERG I.J. et al., "The FAIR guiding principles for scientific data management and stewardship", *Sci. Data*, vol. 3, pp. 1–9, 2016.

[ZHU 15] ZHU Y., LI H., GUO W. et al., "Deciphering genomic underpinnings of quantitative MRI-based radiomic phenotypes of invasive breast carcinoma", *Sci. Rep.*, vol. 5, pp. 1–10, 2015.

Chapter 3

[ASD 18] ASDAL A., GUARINO L., "The Svalbard global seed vault: 10 years – 1 million samples", *Biopreserv. Biobank.*, vol. 16, pp. 391–392, 2018.

[BEN 20] BENECH N., SOKOL H., "Fecal microbiota transplantation in gastrointestinal disorders: Time for precision medicine", *Genome Med.*, vol. 12, pp. 58–62, 2020.

[BRE 18] BREITHOFF E., HARRISON R., "From ark to bank: Extinction, proxies and biocapitals in ex-situ biodiversity conservation practices", *Int. J. Herit. Stud.*, vol. 26, pp. 37–55, 2018.

[CER 16] CERVO S., DE PAOLI P., MESTRONI E. et al., "Drafting biological material transfer agreement: A ready-to-sign model for biobanks and biorepositories", *Int. J. Biol. Markers*, vol. 32, pp. 211–217, 2016.

[CHU 20] CHU DE POITIERS, "Site de la plateforme de recherche morphologique du CHU de Poitiers – Techniques histologiques", available at: https://www.chu-poitiers.fr/specialites/plateforme-recherche-morphologique/techniques-histologiques/, 2020.

[GAI 16] GAIGNAUX A., ASHTON G., COPPOLA D. et al., "A biospecimen proficiency testing program for biobank accreditation: Four years of experience", *Biopreserv. Biobank.*, vol. 14, pp. 429–439, 2016.

[HOF 19] HOFMAN P., DAGHER G., LAURENT-PUIG P. et al., "Tumor banks and complex data management: Current and future challenges", *Ann. Pathol.*, vol. 39, pp. 137–143, 2019.

[ILI 14] ILIE M., HOFMAN V., LONG-MIRA E. et al., "Sentinel circulating tumor cells allow early diagnosis of lung cancer in patients with chronic obstructive pulmonary disease", *PLoS One*, vol. 9, pp. 1–7, 2014.

[KOK 13] KOKKAT T.J., PATEL M.S., MCGARVEY D. et al., "Archived formalin-fixed paraffin-embedded (FFPE) blocks: A valuable underexploited resource for extraction of DNA, RNA, and protein", *Biopreserv. Biobank.*, vol. 11, pp. 101–106, 2013.

[LEE 18] LEE J.E., KIM Y.Y., "How should biobanks prioritize and diversify bio-sample collections? A 40-year scientific publication trend analysis by the type of biosample", *OMICS*, vol. 22, pp. 255–263, 2018.

[MAR 15] MARCHESI J.R., RAVEL J., "The vocabulary of microbiome research: A proposal", *Microbiome*, vol. 3, pp. 31–34, 2015.

[MAR 20] MARQUETTE C.H., BOUTROS J., BENZAQUEN J. et al., "Circulating tumour cells as a potential biomarker for lung cancer screening: A prospective cohort study", *The Lancet*, vol. 8, pp. 709–716, 2020.

[MAS 16] MASCALZONI D., DOVE E.S., RUBINSTEIN Y. et al., "International charter of principles for sharing bio-specimens and data", *Int. Eut. J. Hum. Genet.*, vol. 23, pp. 721–728, 2016.

[MÜL 15] MÜLLER H., DAGHER G., LOIBNER M. et al., "Biobanks for life sciences and personalized medicine: Importance of standardization, biosafety, biosecurity, and data management", *Curr. Opin. Biotech.*, vol. 65, 45–51, 2015.

[NAT 20] NATIONAL CANCER INSTITUTE, "Dictionary of cancer terms", available at: https://www.cancer.gov/publications/dictionaries/cancer-terms/def/biospecimen, 2020.

[SAN 14] SANTIAGO A., PANDA S., MENGELS G. et al., "Processing fecal samples: A step forward for standards in microbial community analysis", *BMC Microbiol.*, vol. 14, pp. 112–121, 2014.

[THE 20a] THE FROZEN ARK PROJECT, "Saving the DNA and viable cells of the world's endangered species", available at: https://www.frozenark.org/, 2020.

[THE 20b] THE HUMAN FOOD PROJECT, "The American Gut Project", available at: http://humanfoodproject.com/americangut/, 2020.

Conclusion

[HOF 20] HOFMAN P., "Challenges and issues surrounding the use for translational research of human samples obtained during the COVID-19 pandemic from lung cancer patients", *Transl. Lung Cancer Res.*, vol. 9, pp. 1543–1553, 2020.

[HOF 21] HOFMAN P., "Gestion des échantillons dans les laboratoires de pathologie et les biobanques au cours de la pandémie liée à la Covid-19", *Rev. Francoph. Lab.*, vol. 528, pp. 36–42, 2021.

[LIG 21] LIGHT E., WIERSMA M., DIVE L. et al., "Biobank networking and globalisation: Perspectives and practices of Australian biobanks", *Aust. Health Rev.*, vol. 45, pp. 214–222, 2021.

[MOU 20] MOUTTHAM L., GARRISON S.J., ARCHER D.L. et al., "A biobank's journey: Implementation of a quality management system and accreditation to ISO 20387", *Biopreserv. Biobank.*, vol. 17, pp. 209–212, 2020.

[RUS 20] RUSH A., CATCHPOOLE D.R., LING R. et al., "Improving academic biobank value and sustainability through an outputs focus", *Value Health.*, vol. 23, pp. 1072–1078, 2020.

[SQA 20] SQALLI H., EL AWADI N., RANCATI C., "Biobankonomics: The sustainability indicators of biobanks", *Med. Sci.*, vol. 36, pp. 282–285, 2020.

[WAS 19] WASHETINE K., HEEKE S., RIBEYRE C. et al., "DNAshell protects DNA stored at room temperature for downstream next-generation sequencing studies", *Biopreserv. Biobank.*, vol. 17, pp. 352–354, 2019.

Index

E, F

G, H

I, L

M, O

P, Q, R

S, T

Other titles from

ISTE

in

Biology and Biomedical Engineering

2024

MASQUIDA Benoît, LECLERC Fabrice
Looking at Ribozymes: Biology of Catalytic RNA
(Nucleic Acids Set – Volume 2)

WALCH Jean-Paul, BLAISE Solange
Phyllotaxis and Symmetry in Angiosperms

2023

BALTHAZART Jacques
Fifty Years of Evolution in Biological Research: Progress and Decline

GILBERT Marc, PIRKMAJER Sergej
Concepts in Biology: A Historical Perspective

VERNA Emeline
Asymptomatic Osseous Variations of the Postcranial Human Skeleton
(Comparative Anatomy and Posture of Animal and Human Set – Volume 5)

WALCH Jean-Paul, BLAISE Solange
Phyllotaxis Models: A Tool for Evolutionary Biologists

2022

DAMBRICOURT MALASSÉ Anne
Embryogeny and Phylogeny of the Human Posture 2: A New Glance at the Future of our Species
(Comparative Anatomy and Posture of Animal and Human Set – Volume 4)

FOSSÉ Philippe
Structures and Functions of Retroviral RNAs: The Multiple Facets of the Retroviral Genome
(Nucleic Acids Set – Volume 1)

2021

DAMBRICOURT MALASSÉ Anne
Embryogeny and Phylogeny of the Human Posture 1: A New Glance at the Future of our Species
(Comparative Anatomy and Posture of Animal and Human Set – Volume 3)

GRANDCOLAS Philippe, MAUREL Marie-Christine
Systematics and the Exploration of Life

HADJOUIS Djillali
The Skull of Quadruped and Bipedal Vertebrates: Variations, Abnormalities and Joint Pathologies
(Comparative Anatomy and Posture of Animal and Human Set – Volume 2)

HULLÉ Maurice, VERNON Philippe
The Terrestrial Macroinvertebrates of the Sub-Antarctic Îles Kerguelen and Île de la Possession

2020

CAZEAU Cyrille
Foot Surgery Viewed Through the Prism of Comparative Anatomy: From Normal to Useful
(Comparative Anatomy and Posture of Animal and Human Set – Volume 1)

DUJON Bernard, PELLETIER Georges
Trajectories of Genetics

2019

BRAND Gérard
Discovering Odors

BUIS Roger
Biology and Mathematics: History and Challenges

2016

CLERC Maureen, BOUGRAIN Laurent, LOTTE Fabien
Brain–Computer Interfaces 1: Foundations and Methods
Brain–Computer Interfaces 2: Technology and Applications

FURGER Christophe
Live Cell Assays: From Research to Health and Regulatory Applications

2015

CLARYSSE Patrick, FRIBOULET Denis
Multi-modality Cardiac Imaging: Processing and Analysis

2014

CHÈZE Laurence
Kinematic Analysis of Human Movement

DAO Tien Tuan, HO BA THO Marie-Christine
Biomechanics of the Musculoskeletal System: Modeling of Data Uncertainty and Knowledge

FANET Hervé
Medical Imaging Based on Magnetic Fields and Ultrasounds

FARIÑAS DEL CERRO Luis, INOUE Katsumi
Logical Modeling of Biological Systems

MIGONNEY Véronique
Biomaterials

TEBBANI Sihem, LOPES Filipa, FILALI Rayen, DUMUR Didier,
PAREAU Dominique
CO_2 Biofixation by Microalgae: Modeling, Estimation and Control